Cambridge Studies in Biological Anthropology 1

Surnames and genetic structure

Cambridge Studies in Biological Anthropology

Series Editors

Surnames and genetic structure

GABRIEL WARD LASKER

Department of Anatomy,
Wayne State University, Detroit, Michigan, USA

With maps and diagrams prepared by
C. G. N. Mascie-Taylor, A. J. Boyce and G. Brush
of the distribution of 100 surnames in
England and Wales

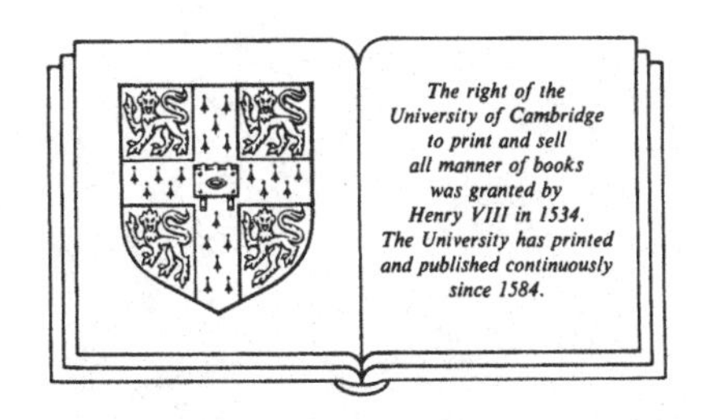

CAMBRIDGE UNIVERSITY PRESS

Cambridge
London New York New Rochelle
Melbourne Sydney

CAMBRIDGE UNIVERSITY PRESS
Cambridge, New York, Melbourne, Madrid, Cape Town, Singapore, São Paulo

Cambridge University Press
The Edinburgh Building, Cambridge CB2 8RU, UK

Published in the United States of America by Cambridge University Press, New York

www.cambridge.org
Information on this title: www.cambridge.org/9780521302852

First published 1985
This digitally printed version 2008

A catalogue record for this publication is available from the British Library

Library of Congress Catalogue Card Number: 84-23827

ISBN 978-0-521-30285-2 hardback
ISBN 978-0-521-05763-9 paperback

Contents

Preface

In 1955 I interrupted my teaching of anatomy for a year in order to teach physical anthropology at the University of Wisconsin. I had already been introduced to the subject of population genetics and at that time occasionally met with and discussed the subject with J. F. Crow and N. E. Morton. Back in Michigan the members of the Department of Human Genetics at the University of Michigan and other geneticists continued to stimulate these interests of mine. However, some of the mathematical models, which are common in population genetics, seem abstract and in my research I have preferred more comprehensible (even if somewhat less elegant) formulations when they are adequate to explain the empirical data. About two years ago, G. A. Harrison of the Department of Biological Anthropology at the University of Oxford expressed his pleasant surprise at how much interesting information on the structure of human populations had been garnered by studies of the distribution of surnames, and he suggested that I write a little book on the genetic structure of human populations as seen in this way. He implied that such a book should be directed at an audience who may not be at home with the erudite algebra of population genetics. I have therefore avoided discussion of some minor theoretical differences between mathematical models. Those who wish to pursue such issues will find additional definitions in the Glossary and appropriate references cited in the text. However, random variation plus exceptions to the assumptions account for more of the deviation from exact fit of surname models to observable facts than do these differences in theory. For this reason, when one deals with actual data one usually has to shift the focus from general theory to specific elements in the cultural and genetic determinants of the demography of the population in question. The problem is thus anthropological.

An anthropologist is as concerned with non-genetic influences as with genetic ones. If a work on a primarily genetic topic is enhanced by this stance and contributes to the sorting out of both kinds of influence on population structure, I have to thank anthropologist colleagues, especially my wife, B. A. Kaplan, for their influence on my thinking. She and others have worked with me on specific aspects of the topic of this book and they are cited where these aspects are discussed. Among these collaborators, I am especially grateful to C. G. N. Mascie-Taylor, who sponsored us for fellow commonerships at Churchill College, Cambridge, and I thank him and his colleagues in the Department of Physical Anthropology, University of Cambridge, for facilities for our work and an agreeable and stimulating environment.

I also thank the National Science Foundation for support for some of my studies of the population structure of the British population, and the Wenner-Gren Foundation for Anthropological Research and its director of research, Lita Osmundsen, for their sustained interest in these studies and for a grant for this work. The American Philosophical Society also contributed support.

J. L. Boldsen has written computer programs for some still unfinished extensions of these studies. D. F. Roberts, C. G. N. Mascie-Taylor, A. J. Boyce and B. A. Kaplan criticized a draft manuscript.

G. W. L.

Cambridge, England

1 *Introduction*

The application of models based on surnames to the study of the genetic structure of the human population would seem to call for some justification. Any such application involves the assumption that the inheritance of surnames and biological inheritance are similar, or alternatively it must attempt to measure and allow for the differences between the inheritance of surnames and that of genetic traits.

One may begin to introduce the subject of surname models by an account of the scope of human biology, the place of human population structure in it, and the reason that models by analogy are needed. Human biology is concerned with the adaptive mechanism that makes human life possible. From one point of view this is controlled by those aspects of the genome that are shared by all humans and distinguish human beings from members of other animal species. Human life involves the response of human beings in various cultural and natural environments.

The other chief concern of human biology is human differences and the factors that account for them. Again these can be genetic at base, but they also involve interaction with the environment – which, for human beings, involves virtually the whole range of land habitats and is rendered much more varied still by the results of human activities and their variation from region to region.

The whole question of human biology, with complex diversity within the species overshadowed by the similarities among humans which distinguish *Homo sapiens* from other species, can be sketched synthetically but it cannot be studied as a whole. Instead, specific problems have to be isolated and attacked one at a time.

The landmark studies of human biology are of this kind. For instance, the classic report to a US Senate immigration committee by Franz Boas (dated 1910, but almost always cited as 1911) appeared at a time when the cephalic index was considered to be a hallmark of race and hence inherent and immutable. However, Boas showed that children reared to adulthood in a different country from that in which their parents were reared, and hence in a different environment, grow up to have, on average, a different cephalic index and a different stature. Boas' findings have since been amply confirmed for Jews, Mexicans, Japanese, Chinese and other immigrant groups in the United States.

Another result which has proved to hold true in many subsequent investigations is Raymond Pearl's finding that tobacco smokers tend to live less long. Pearl had been associated with the eugenics movement and believed in a predominant role of genetics in human experience, including the determination of length of life. Probably it is no accident that someone with such a slant – that is, the belief that longevity is genetically determined – would successfully demonstrate the importance of an environmental factor. Those who need no convincing of the importance of environmental influences, on the other hand, may be expected to impose rigorous controls on them and produce good studies of the human genetics of quantitative traits.

All human biological attributes have both genetic and non-genetic aspects, of course, but that is not to say that specific observable differences are always the result of combined influences. In studying surname models, for instance, one is concerned purely with the genetic. Surnames do have non-genetic factors associated with their origin and mutation. These are peculiar to surnames, however, and are not like the non-genetic influences in biological mutation. That is, the origins of surnames are to be found in linguistic phenomena similar to those of other kinds of naming such as place naming. The rules for changing personal names in adoption, on marriage, or for other reasons are purely cultural rules. They must be understood and allowed for, not to include them in the models, but to minimise their influence on the models.

Illegitimacy has sometimes been cited as a condition in which the child has a different surname from its father and hence in which the assumptions in surname models of genetic transmission would be violated. In many countries, including England, however, the illegitimate child of an unmarried mother usually takes the mother's surname. If the purpose of modelling is to use the line of descent marked by surname as representative of all lines of descent (as is the case in the use of marriages between persons with the same surname for estimates of inbreeding) then lines passing through mothers are as valid as those passing through fathers, and children taking their mothers' surnames pose no problems. Illegitimate children of married mothers who take the surnames of their mothers' husbands do produce errors in the surname models, but the same instances usually cause the same problem in direct studies of the pedigrees. In recent years people in Western society have become more open in discussion of these issues of paternity and, by intensive interviewing, it might now be possible to estimate how many cases of 'wrong' surnames are being included in a survey. In this way one could also recalculate coefficients derived from the survey to see how much difference these false classifications make. Until now, however, users of surname models have merely hoped and believed that, on average, the mothers' husbands and biological fathers are similar enough in relevant characteristics indexed by surnames (such as ethnic and geographic origins and consanguinity with the mother) for a few instances of mistaken paternity to have no significance.

One peculiarity of surnames is that there tend to be sharp discontinuities in the occurrence of particular surnames at national and other linguistic boundaries. These discontinuities may be sharper than are found in gene frequency distributions because, in the south and west parts of the European continent, present distributions of surname frequencies mark the results of migrations of approximately the last 40 generations whereas gene frequency distributions are the result of the ebb and flow of migrations over a very much longer, although indefinite, past. Thus the greater geographic variability of surnames than of human genes over the whole continent is not entirely – and perhaps not even importantly – due to the fact that pronunciation (and hence the way surnames are spelt) may be modified by local usage whereas genes remain in unaltered form among migrants and their descendants.

Surname models are important, then, for the very fact that they isolate genetic aspects and deal with them separately. The more one believes genetic–environmental interaction to be important in human biology, the more reason there is to start one's analysis in situations in which one or the other factor is minimized. One would stand to be criticized for this only if one considered human population genetics to be the whole of the science of human biology.

However, this is one aspect of surname models that has been criticized: what they do not do. Even within the genetic sphere, surname studies have not been (and cannot be) directly applied to the evaluation of natural selection. Natural selection is a mechanism by which genetic–environmental interactions influence genetic consequences, and the purely genetic approach of surname models is bound to miss this subject just as it misses questions of direct environmental impacts on human biology of the kinds posed by Boas and Pearl. Weiss and Chakraborty (1982) said: 'the selection–drift controversy is the central problem in evolutionary population genetics today. This is largely because there is no adequate selection model to explain the maintenance of the vast amount of polymorphism that has been found.' There are so many different specific distribution patterns of known human polymorphisms that no single or few modes of selection would account for them all and some investigators attribute them to stochastic (chance) factors. Others think that random or stochastic merely means so far unexplained, but that all the patterns have their *raison d'être*. Weiss and Chakraborty, who seem to be speaking from the first of these positions, complain that the physical anthropologists who have studied human inbreeding by means of the prevalence of marriages between persons of the same surname, have pursued their studies with little sense of this problem. Indeed, the whole of the anthropological effort to study the breeding structure of human populations and of the species as a whole could be characterized as a descriptive enterprise.

This misses a main point of such studies, however, for there is an implicit programme in the study of surname genetics. Precisely because surnames cannot be subject to the forces of natural selection and must be considered neutral to selection by disease or climate, a difference (or lack of it) in geographic patterns of the tens of thousands of surname 'alleles' compared with those of biological alleles would be evidence bearing on what Weiss and Chakraborty call the central problem in evolutionary population genetics today: the selection–drift controversy. Unfortunately one cannot conclude that all differences between findings from surname distributions and those from studies of polymorphic genes are due to the action of selection in the latter case but absent in the former. It is also necessary to allow for the indefinite time span of biological genes and the rather limited history of surnames. This difference in itself permits approach through surname analysis to another important problem: the tempo of changes. Such estimates of the rate of change may be derived from surnames of several centuries' antiquity but be difficult from pedigrees of only three or four generations' depth and impossible in the equilibrium state reached by genes after an indefinite span of thousands of generations.

There are still other complications with surname analyses: the high frequency of multiple independent origins and mutations. Such shortcomings are not unique to this type of study; they are to some extent present in the alternative methods of approach. Even if the descriptive and historical information gained by surname analysis is not valued for itself, the light shed on systematic versus random selection warrants the addition of surname analysis (a relatively small additional effort) to other methods of study in human population genetics.

Whatever the merits of this line of thinking, the mere description of present and past population structure has value because of the political misuse to which faulty concepts can be put. History is fraught with disasters caused by the misconception that the human population is formed of pure races and recent mixtures between them. The alternative conception, promoted by some extreme environmentalists, of a formless mish-mash, may be damaging too, because its rejection by the populace on the basis of their own experience may add fuel to, rather than dampen, the appeal of racialist views.

These various reasons may help justify the pursuit of surname models of population structure. Surname studies are by no means the end-all of human biology, which must involve the understanding of both genetic and environmental components of human variation and the interactions between them. However, surname genetics has a certain fascination which it is hoped will be conveyed by its explicit and implicit applications to questions that go beyond the origin, spread and extinction of surnames themselves. It is that fascination and the ready availability of research material that has led to the development of the body of knowledge on the subject.

Surnames are not distributed homogeneously in different places and among different social groups. The general purpose of surname studies in human biology is to measure the different probabilities of finding the same surnames in different times, places, groups and, especially, in marital partners. These probabilities can be compared with gene frequency distributions and assortative mating for genes of polymorphic systems. Similarities will allow one to use surnames to model the genes; differences may help in understanding the processes of differentiation of gene frequencies.

2 *History of surname studies in human biology*

Yasuda and Morton (1967) traced the history of the use of surname models for the study of human inbreeding to George Darwin's (1875) article in the *Journal of the Statistical Society*. Darwin's father, the famous naturalist Charles Darwin, and his mother, a member of the Wedgwood family of china pottery fame, were first cousins. Darwin was interested in the possible deleterious effects of consanguinity of parents and he wanted to know the frequency of cousin marriages in England. He therefore sought data on cousin marriages and on marriages between persons of the same surname in various sources such as *Burke's Peerage* and the Pall Mall social register. He then followed an ingenious line of thinking to estimate the proportion of marriages between first cousins. He reasoned that marriages to a person of the same surname who was not a first cousin would be proportional to the frequency of the surname in the population. This would be frequent only for common surnames. The Registrar General (1853) had published the frequency of the 50 most common surnames in the marriage registers and from the sum of the squares of these frequencies (0.0009207) Darwin estimated that marriages between unrelated persons of the same surname would be not much different from one per thousand. The excess over this of marriages of persons of the same surname was ascribed to cousin marriages and this was divided by the fraction of cousin marriages that were same-name marriages to give the number of cousin marriages in the population. Darwin concluded that the rate of first cousin marriages was about 4½% among the aristocracy, 3½% among the middle classes and landed gentry, 2¼% in the general population of rural districts and 2% in the cities.

Thirty-three years later, Arner (1908) published a similar study of cousin marriages in eighteenth-century New York and nineteenth-century Ohio. He first examined 10198 marriage licences in New York dated before 1784. The 50 commonest surnames gave an expectation of same-name marriages at random of 0.000757, but 211 of the marriages (0.0207) were actually between partners of the same surname. Using Darwin's ratio of same-name to all first cousin marriages would have led to an estimate of 5.9% cousin marriages in colonial New York. Arner considered this estimate to be too high. He found genealogies in which there were 242 same-name marriages of which 70 were between first cousins. He believed that this rate might be biased by pedigrees in which only males were traced so he eliminated the all-male pedigrees and found that 24 of the remaining 62 same-name couples were first cousins. After rounding the ratios, Arner calculated that 2.76% of the marriages in colonial New York were between first cousins. In Ashtabula County, Ohio, he found records of 13309 marriages between 1811 and 1886 of which 112 were between persons of the same surname. He concluded from this, by Darwin's method, that 1.12% of these marriages were between first cousins.

Darwin's and Arner's studies were not immediately followed by others and they were apparently unknown to those who first revived study of the subject. In 1964 or so James F. Crow was invited to write an article on human inbreeding to accompany publication of one by Gordon Allen on random and non-random inbreeding. Crow recalled that in a lecture in the 1940s H. J. Muller (who received the Nobel Prize for his discovery that X-irradiation stimulates genetic mutations) had suggested that surnames could be used in genetic models of inbreeding. Crow (1983) says that he then had the thought for which his work on surnames is usually cited – that in most relationships the inverse of the likelihood of having the same surname times the degree of relationship is a constant number. Offspring of a brother and sister have an inbreeding coefficient of $\frac{1}{4}$ and always bear the same surname. Offspring of aunts and uncles with their nephews and nieces have an inbreeding coefficient of $\frac{1}{8}$ and bear the same surname in approximately one half of the instances, thus also yielding, on average, $\frac{1}{4}$ (that is $\frac{1}{8} \times \frac{2}{1}$) to the inbreeding coefficient of the population. Offspring of first cousin marriages have an inbreeding coefficient of $\frac{1}{16}$ and one type of cousin in four bears the same surname (fathers' brothers offspring, but not the offspring of fathers' sisters, mothers' brothers or mothers' sisters); thus the contribution to the inbreeding coefficient of the population represented by each married pair of same-surname first cousins is $\frac{1}{16} \times \frac{4}{1}$, which is also $\frac{1}{4}$. Crow took his idea and set forth this reasoning to a colleague, Charles Cotterman, who then worked on the problem of the possible relationships among four people and concluded that there are 17 such relationships and that 13 of them fail to agree with the 1 : 4 ratio. That is because across generations one's mother has a different (premarital) surname from oneself. Because of this theoretical difficulty Cotterman did not join Crow in writing the paper. Crow saw, however, that most consanguineous marriages are not across generations and that the four instances of the 1 : 4 ratio on Cotterman's list encompass most of the instances in human marriages and hence in human mating. Crow also wrote to Muller to tell him about this development of his idea, but to Crow's astonishment Muller had completely forgotten and said he had never heard of such a thing.

Crow was interested in two prospects for the use of surname models: first, inclusion of remote common ancestry that would be revealed by a surname in common but missed in pedigrees from interviews; and second, separating the non-random component of inbreeding from the random one. The random component is the amount of inbreeding due to the limited size of the population breeding as though all possible matings had been equally probable. The non-random component is the extent to which this is increased (or decreased) by selective mating in a single generation. In order to work through these ideas on suitable data, Crow sought the cooperation of Arthur Mange, who had assembled records of the marriages of the Hutterites, a religious isolate living in the western United States and Canada. The Hutterites had collected much information about themselves and were willing to cooperate with Mange and other geneticists. Cotterman contributed further advice and Crow and Mange (1965) wrote about them in the most influential article ever to appear on the subject of surname models of human inbreeding. In it they estimated the inbreeding coefficient (in a population, the average proportion of genes at paired loci that are identical by descent from the same ancestors through both parents). Crow and Mange took the rate of inbreeding to be one quarter the frequency of marriages between persons of the same surname, which they called 'marital isonymy'. Furthermore, they partitioned the inbreeding coefficient into a random and a non-random component by a method they developed which adapts the approach of Wright (1922).

The idea of using surnames to study inbreeding was not new with Crow and Mange's publication. Although Crow and Mange were unaware of it, Yasuda and Morton (1967) knew of the studies by Darwin (1875) and Arner (1908) and that an American geneticist, Shaw (1960), had also pointed out that regular use of two surnames per person in Spanish-speaking countries provides an opportunity for effectively applying an index of consanguinity. In the Spanish system the given name or names is followed by the father's surname and then the mother's father's surname. Thus the last name is dropped each generation and replaced by a new name – that of the mother's father. Since married women usually retain their maiden names, this identifies an individual with both parents' families orientation (i.e. the families into which their parents were born and in which they grew up).

In the meantime (according to Yasuda, 1983), and without knowledge of the inbreeding coefficient developed by Wright (1922), Kamizaki (1954) had calculated the expected frequency of isonymy (i.e. having a surname in common) among various degrees of relatives (as Cotterman and Crow were to do later) and anticipated results of Crow and Mange. Kamizaki also cited earlier Japanese work in which the proportion of consanguineous marriages was estimated from isonymy. He reported too the same proportions of isonymy in consanguineous marriages as later derived by Crow ($\frac{1}{4}$ in first cousins, $\frac{1}{8}$ in first cousins once removed, $\frac{1}{16}$ in full second cousins, etc.) and derived general formulations for estimating the degree of relationship due to more remote consanguinity. Yasuda points out that change of a man's surname by adoption of his wife's, a frequent occurrence in Japan, will not ordinarily change the average frequency of isonymy. That is, it will shift the tested relationships from all-male lines to lines with female links, but will retain the same level of probability. This conclusion does not apply to the usual type of adoption in Western society, but does apply to cases of illegitimacy where the mother's surname is used for the child.

One other point should be made about work with Japanese surnames, however. Prior to the Meiji restoration, surnames were not allowed to be used except by the governing classes, and did not become mandatory until 1875. So it is only slightly over a century since surnames were arbitrarily assigned to almost all founders of the present surname lines. Thus, in Japan, inbreeding calculated from isonymy is for a period no more than about five generations – a length of time that can be encompassed by careful interviews and searches of family and other records. Since pedigrees include all consanguineous unions, but isonymy counts only a fraction and estimates the rest, over this span of time pedigrees provide the better way of determining the extent of inbreeding. On the other hand, the very fact that the time span of the use of surnames is about the same as that which can be covered in pedigree studies greatly enhances the value of a direct comparison of results by the two methods for evaluating the applicability of isonymy levels to estimating inbreeding.

Other studies in which several methods have been applied (among them studies in Switzerland and elsewhere by Morton and associates and by Ellis and associates, in the Pyrenees by Bourgoin and Vu Tien Khang and in the Orkney Islands by Roberts and Roberts) show rather poor correspondence of estimates of inbreeding from surnames and from pedigrees. In the West, unlike Japan, surnames have a considerable antiquity and the higher estimates of inbreeding from surnames than from pedigrees in some of these cases are at least partly explained by the inclusion of remote inbreeding by the surname but not by the pedigree method.

Communities where there are few surnames some of which occur with high frequency must always have been known to be inbred. In 1957–8, during a comparative study of a number of communities on the north coast of Peru, I was aware of this and collected lists of surnames from various sources (interviews, birth registers, death registers and grave markers) and for various periods of time so that a scale of isolation or inbreeding could be devised and compared with rates of migration into the same communities. When it came to analysing the data, however, it was not clear how one could deal with the surname distribution data. Only much later, Wendy Fox, a mathematician, suggested that one could approximate the frequency distribution curve of surnames with a formulation that would permit comparison of the constants of curves fitted to different sets of data (Fox and Lasker, 1983). If the surnames of a population are listed by rank order of their frequency of occurrence, the log of number of surnames occurring x times against x tends to form a straight line with the log of number of occurrences. According to the reasoning of Zipf (1949) this would be so if the rate of growth or decline in frequency of a surname were independent of its frequency. The formula gives a good fit to some data from several adjacent areas in Berkshire and southern Oxfordshire (from marriages registered in Reading, Wokingham and Henley). Unfortunately the slopes of lines fitted in this way and of similar curves that can be fitted to surname frequency distributions are dependent on sample size. That had been apparent in the sets of data from Peruvian towns and villages. When the Crow and Mange (1965) article appeared a method was available for application to them that is unaffected by sample size (Lasker, 1968, 1969).

Recently, Zei *et al.* (1983*a*,*b*) have argued that the theoretical distribution of neutral alleles – the assumed model for surnames – is better matched by a logarithmic distribution originally introduced by R. A. Fisher to represent the variation in the abundance of species and applied to surname frequencies by Chakraborty *et al.* (1981). Zei *et al.* (1983*b*) compared the fit of these different formulations and in some examples found their method seems to describe the data more precisely than the method of Fox and Lasker. Furthermore, they showed how to take sample size into account and published a table for use in the necessary computations. Wijsman *et al.* (1984) have further extended these studies by developing a method for calculation of migration rates from the matrices of surname frequencies in the various places in an area at two or more periods of time.

The history of surname models does have a summary chapter. At the meeting of the American Anthropological Association in 1980, K. Gottlieb, who had been conducting a study of the population of the Island of Sark in the English Channel, suggested that a conference and symposia be held to bring together those who were using models based on surnames to study the breeding patterns of human populations. One theme of the conference was the use of surnames as markers of the populations in which they occur. Azevedo described her studies of surnames as indicators of racial origins in Brazilian cities of mixed racial composition (Azevedo *et al.*, 1983). Stevenson *et al.* (1983) used surnames to sort out the European countries of origin of residents of a North American city studied with respect to genetic polymorphisms. And Gottlieb (1983) explored the value of Spanish surnames as markers of Mexican ancestry in the United States. Another group of contributions dealt with inbreeding. Swedlund and Boyce (1983) described the population structure of the colonial population of the Connecticut River Valley and others drew examples from islands in the West Indies, Scotland, and the coast of Maine. Yet other contributors dealt with the study of subpopulations within a larger region (Zei *et al.*, 1983*a*; Chen and Cavalli-Sforza, 1983; Hurd, 1983) and with the tendency of rare surnames to be more highly localized than common ones (Lasker, 1983; Zei *et al.*, 1983*a*).

The difficulties with surname models were not ignored. Roberts and Roberts (1983) used data from the Orkney Islands to show how little connection there may be between inbreeding estimated by isonymy and inbreeding in the same individuals calculated from pedigrees. K. Weiss and his associates (1983) dealt with another of the difficulties with using surnames in genetic models: the occurrence of changes (mutations) in the names. Thus some of the contributors were critical of the use of surnames in biological models, but, as might be expected from a group who had been involved in such studies, the general response was enthusiasm for research of this kind.

3 *Sources of data*

Well over a million different surnames exist. This is true despite the fact that the largest society, that of China, has only a few hundred surnames and there are still some societies with none. The system of hereditary surnames (usually passed from father to offspring of both sexes – but of equal value in surname models where some other system prevails) can be thought of as a gene with over a million alleles. With that many alleles the system is much richer in informational content than any biological gene. The HLA system of tissue compatibility antigens, which is the most variable human genetic system so far investigated, has about four orders of magnitude fewer known variants than the system of surnames.

The first method to be considered in the present context is the mere counting of surnames. How a surname will be defined for this depends on the use that is to be made of the results. To the extent that the purpose is to simulate genetics, there is no point in grouping together two surnames that may have had their origin from the same profession (Mueller, Miller), personal attribute (Blanc, Bianco, White, Weiss) or saint's name (Martinez, Martin, Martini) if they were originally attached to unrelated families. On the other hand, since one is interested in considering together families that stem as branches from the same trunk, one may wish to merge two different surnames that derive from the same place name (Rotherham, Rudram). There were no standard spellings until they were imposed by national academies, dictionaries, encyclopaedias and other institutions of the last three centuries. In earlier renditions of surnames (and even occasionally recent versions) consonants were doubled or left single, similar consonants were substituted for each other, endings varied, and vowels changed at the whim of the scribes who prepared the documents. Thus, Mitchel and Mitchell, Cane and Kane, York and Yorke, Rowes and Rose, and Read, Reed and Reid were merely variable ways of writing the same surnames.

One may well be justified in grouping together those names thought to be the same – especially if one is concerned with an early common origin. However, in recent times sons and daughters of a Reed usually spell their name 'Reed', not 'Read' or 'Reid', and so forth, so if the chief concern is with recent common origins, it would be well not to merge such spellings. Whether to merge variants of spelling or not may make a considerable difference to results. Buchanan and his associates (1982) have produced a huge list of equivalences of the Spanish surnames that occur in Laredo, Texas. Equivalences are no doubt even more numerous in England since standardization of spelling is more recent and more informal in English than in Spanish. Under the circumstances, a sensible strategy in a surname study is to conduct the analysis with each spelling considered to be a separate entity, and then to do the study again after variant spellings have been merged (Küchemann *et al.*, 1979). In a study of early eighteenth-century household census entries in villages of East Kent, Souden also tried comparing results after using more and less stringent criteria for merging names. The numerical results were appreciably different in the two analyses, but the rank order of results from the different villages was much the same, as were general findings about geographic distributions (Souden and Lasker, 1978).

After the issue of what constitutes a surname has been decided, the next issues are what constitutes a population and how to sample it. If one is modelling genetic processes, the population should be defined in the same way as human geneticists define it. This involves some circular reasoning since some genetics textbooks define the gene pool as the totality of genetic information of a breeding population, and the breeding population as those individuals who share the same gene pool. To avoid the dilemma of defining who breeds with whom in this circular way, one should use social criteria and define the breeding population as the reproducing members of a community. In large complex societies it may not be clear exactly who belongs to what community, however, and it is sometimes necessary initially to use arbitrary criteria such as political divisions, parishes or whatever units are available in the data sets. Such arbitrary units can be combined or redefined later after the data have been analysed with respect to mate selection.

In regional studies of marital migration, such as that of Hiorns and his colleagues (1969) of the Otmoor villages and Coleman's (1979, 1980*a*) study of the city of Reading and its surroundings, there may be a category 'outside world' and, because the pattern of male migration is different from that of female migration, studies of male and female surnames of such a category would differ considerably (Lasker *et al.*, 1979). Male migrants come in larger numbers than female and from more distant places; therefore males in such a category bear more of the more exotic surnames that were formerly absent locally. Other arbitrary groupings besides 'outside world' may also not correspond with real social units and this may lead to similar, although usually less dramatic divergencies of the defined populations from the true breeding population units. On the other hand, islands and island groups generally have natural geographic boundaries to their human populations and subpopulations. Religious isolates also have sharp breeding boundaries because, although some members may leave the group, few, if any, outsiders marry in. Isolates sometimes are divided into populations called *leut*, the term used by the Hutterites for the major historical groups of their fissioning colonies.

Another important consideration is that because of the enormous number of different surnames, samples adequate for studying them separately have to be of large size. Data on small populations, even if these are sampled completely, may be too few for useful conclusions unless a number of similar small populations are studied and generalizations are based on the findings from several of them rather than on specific communities. Since surname methods are usually employed to derive general statements about population structure rather than to reconstruct local population history, such an accumulation of small studies is an entirely appropriate strategy.

Any random or complete list of surnames can be used for surname analyses. Each type of source has advantages and disadvantages. Records of births, deaths and marriages are possibilities. One advantage of birth records is that in past times births ordinarily occurred in the home and were registered in the parish or registration centre to which they pertained. Since births in hospital became common, however, registration may be at some distance from the residence. Even now, though, the place of residence is often listed and can be used instead of the place of registration if original documents or direct copies are available for study and a high degree of specificity is required. In the nineteenth century and earlier, infant death rates were high, so that a list of births would include many individuals who were never to join the breeding population. Baptismal records are similar to birth records and have some of the same advantages and disadvantages. In addition, if there are several religious groups in the population, such as Church of England, Nonconformist, Roman Catholic, Quaker and Jewish, only members of one religion may be listed; or if all groups are required to list vital events in the church of the official religion, compliance may be incomplete in the case of certain sects. Records may also be incomplete with respect to females, and, if there is a fee or tax or obligation for military service or other duties for which the authorities might use the vital listings, some males may not be registered.

Lists of deaths have many of the same drawbacks as those of births, but they also have some of their own. In most societies married women are listed by their husband's rather than their father's surname so the genetic analogy is much less direct than for lists from some other sources. Also, because people die at different ages, the question arises as to the breeding population of what period of time it is that is being sampled. By contrast with birth records, death records do not yield like-aged cohorts. Like birth records, however, those who die in infancy and are never to enter the breeding population will be counted (unless they are individually excluded by reference to other information on the death records such as age at death or marital status). Whereas the parents usually give the information for birth records and the individuals involved give the information for marriage records, data on death records may come from offspring, more distant relatives or totally unrelated individuals, and are sometimes inaccurate. Death records can be checked, supplemented or substituted for by data from inscriptions on grave markers, but not every grave has a marker and the inscriptions vary in durability; the more permanent ones in stone and brass are likely to be for individuals of high social status and a study of surnames from such markers would be biased towards representing a higher than average social position.

Marriage records provide an excellent source of surname data for genetic analysis. Virtually all members of the breeding population are married and hence included. Brides are listed by their maiden names and the names of the parents of brides and bridegrooms can be associated to determine the level of marital isonymy. Furthermore any tendency of specific surnames to be associated with each other in marriage more often than would be expected by chance (as would be the case in marital exchange between families, or in populations subdivided into ethnic groups, moieties, clans and other structures) can be studied by analysis of the matrix of brides' surnames by bridegrooms' surnames. One shortcoming of marriage records is that they may not always list places of residence or birth. Since marriages usually take place at or near the residence of the partner of one sex (in European tradition, the bride), the population to which the partner of the other sex belongs may be inaccurately specified. Another problem of recently increased frequency is encountered with couples who have already established a union before marriage in which the woman may have called herself by her future husband's surname. Second and subsequent marriages also pose a potential problem since brides may be listed by the surname of former husbands and individuals of both sexes will be sampled more than once by the marriage records.

Lastly, ways exist to use censuses of one sort or another. Raw national census data which include people's names are usually not made available for research for 70 or 100 years, so cannot be used for modern periods. Censuses, like birth and death records, include children, but it is often easy to limit analysis to individuals of a specific age range and marital status. In census data married women are usually listed by their husband's surnames, so it may be desirable to limit analysis to one sex or the other. Studies of adults from household census data that include both sexes will contain far more surnames that occur twice (relative to the number that occur once only) than would a study of the same size but limited to one sex. Special censuses, such as registries of eligible voters, are a useful source of information about the surnames of the current population. Directories, including telephone directories, billing addresses for public utilities or department stores, and other lists provide information on surnames of family units that can be associated with locale and used for surname research. In developed countries, such as England and Japan, lists of these kinds have yielded results similar to those from more universal types of samples; that is, in recent times special lists of these kinds seem not to be noticeably biased in respect of surnames.

All sources of data have some shortcomings, but the limitations are different for different sources. Therefore the problems can be circumvented by using data from more than one type of source and comparing the results. Few studies have done this, but it is reassuring that those that have, report similar results from samples from diverse kinds of sources (e.g. Lasker, 1969).

4 *Methods*

Surname frequency distributions

The simplest use that can be made of surname data in human biology is to record and compare frequency distributions. If the surnames of all the members of a population (or of a random sample of it) are arranged in rank order from the most to the least frequent, a series of statements can be made about such matters as the frequency of the most frequent name (or several of the most frequent names). Statements can also be made about rare names, for instance the number of names that occur only once. If the samples are large, the frequency of each common surname should vary only by sampling error and should not be appreciably affected by sample size. The number and proportion of surnames that occur once and only once, however, are very dependent on the sample size: as one begins sampling, every name will be unique, but, since the number of surnames is finite, as one continues sampling one will approach a condition in which all surnames in the sample have been encountered a second time. The same applies, although less markedly, to surnames that are listed only twice or three times and to other rare surnames. Thus the shape of the curve of the distribution of surname occurrences is dependent on sample size. In comparative studies a very simple solution would be to draw samples of equal size, but I do not think that this has yet been done in surname studies.

Dobson and Roberts (1971) used a method introduced by Buckatzsch (1951), in which the proportion of surnames surviving from a prior period of time is taken as a measure of population constancy. This measure is also subject to the size of the samples. Thus if the population is completely sampled at two periods of time and its size has increased or decreased substantially in the meantime, such a measure of constancy will be affected. This is manifest and explained in a study of a rapidly expanding population in a Tasmanian census district (Kosten *et al.*, 1983).

The approach to comparing distributions can be shifted from strategies which emphasize only common or only rare surnames by dealing instead with the whole distribution. If such distributions regularly follow a characteristic curve, then any sample can be characterized by the slope and intercept of the straight line transform of the curve. The problem is similar to that of the frequency of the different words in a text, as discussed by the mathematician Zipf (1949). Yasuda and Saitou (1984) fitted straight lines to the log of occurrences versus the log of ranks in the frequency distributions given in ranked lists of surnames. Likewise, Fox fitted straight lines to the log of the number of surnames occurring x times plotted against the log of x (Fox and Lasker, 1983). Similar formulations have been set forth in publications by Yasuda *et al.* (1974), Zei *et al.* (1983*b*) and Yasuda and Saitou (1984). A rank order curve of surname occurrences can be fitted reasonably well by any of several formulations, but the task is to choose one that can be shown to result from the forces that presumably make for the distribution as implied by Zipf and claimed for the Karlin–McGregor formula by Zei *et al.* (1983*b*).

One method for comparing surname frequency curves is to find the number, x, of people with each surname in a series of equal-sized random samples, and from the rank-ordered list of x, list $f(x)$, the number of surnames that occur x times. According to the discrete Pareto distribution, the probability that a surname occurs x times, $P(x)$, is proportional to $x^{-(c+x)}$, where c is a positive constant. Values of c can be determined by the maximum likelihood method.

By fitting part of such a curve it might be possible to project the rest and apply the constants to such problems as occur when the frequency of common surnames has been counted but the frequency of rare ones is unknown. Such a projection must be interpreted with caution, however: the part of the curve determined by common surnames, although the part most likely to be known, has considerable stochastic variability in slope and is of least value in genetic models because these are the very surnames most likely to be polyphyletic.

For simple analysis of samples of standard size, the number of different surnames, k, may be used. This gives considerable information about the frequency distribution but saves the labour of curve-fitting. Unlike the constant c, k is largely determined by rare surnames, the ones with the most genetic significance because they are the least likely to be polyphyletic. Some studies have reported values of k, but, unfortunately, if the sample sizes vary, direct comparison of k or of $k/\Sigma f(x)$ is not in itself meaningful. With samples of equal and adequate size, within-population variation in k would be small relative to $\Sigma f(x)$ and significantly different population structures would be expected to yield significantly different values of k. Such a comparison would only apply to the internal structure of the populations, of course, because two populations with identical values of k (or identical values of c, for that matter) might consist of persons with totally different surnames. For this reason, methods for comparing information on each surname separately yield further information not present in name frequency data alone.

Isonymy, relationship and inbreeding

Isonymy simply means possession of the same surname. The proportion of isonymy is the frequency with which this occurs: interpopulation isonymy is that between two samples, marital isonymy that between a bride (using her maiden name) and a bridegroom. The term isonymy is sometimes limited to marital isonymy or used as an estimate of inbreeding from the proportion of isonymy, but such limitations and extensions may be confusing and the term should not be used in these ways without an explanation. Fig. 1 shows how intrapopulation and interpopulation isonymy are calculated.

Geneticists have used the terms relationship, kinship, consanguinity and inbreeding with different meanings (Jacquard, 1975). Here the term 'inbreeding' is taken to mean the probability that two genes at an autosomal locus are identical by descent through the mother and father from the same ancestor. For a population, the inbreeding coefficient is the average of this probability among its members. The genetic relationship

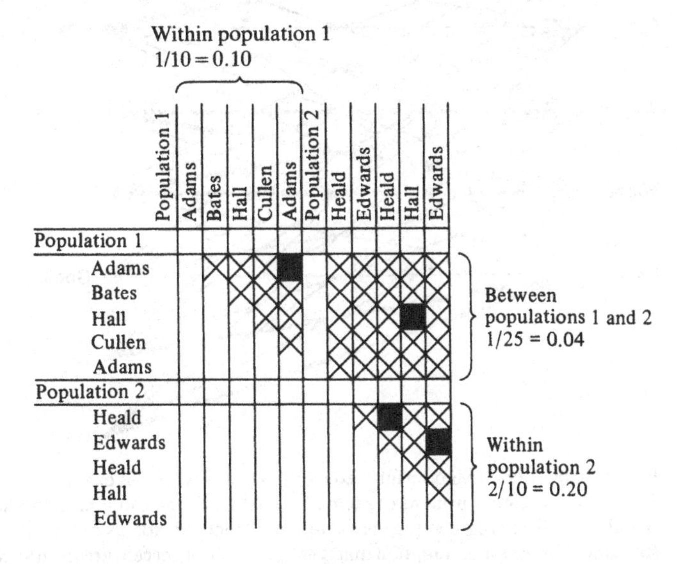

Fig. 1. Within-population and between-population isonymy. Black squares represent isonymous pairs; crosses represent all other possible pairs.

is the probability that two individuals share an autosomal gene by reason of descent from the same ancestor. Kinship is reckoned as twice the degree of relationship, but the degree of inbreeding in an individual is just half the degree of relationship of the parents to each other. For example, in our society a brother and sister bear the same surname and the rate of isonymy between brothers and sisters is therefore 1.0. Brothers and sisters share half their autosomal genes so their genetic relationship is 0.5. If a brother and sister mated, offspring would, by reason of this common descent, be expected to have the same gene at one quarter of their homologous autosomal gene loci; hence their inbreeding coefficient would be 0.25. Under the assumptions about surnames stated above, therefore, when a proportion of isonymy is expressed as a genetic coefficient of relationship by isonymy (*Ri*), it is divided by 2, and when it is expressed as an inbreeding coefficient by isonymy, it is divided by 4. The method of calculating the random component of the inbreeding coefficient by isonymy is shown diagrammatically in Fig. 2.

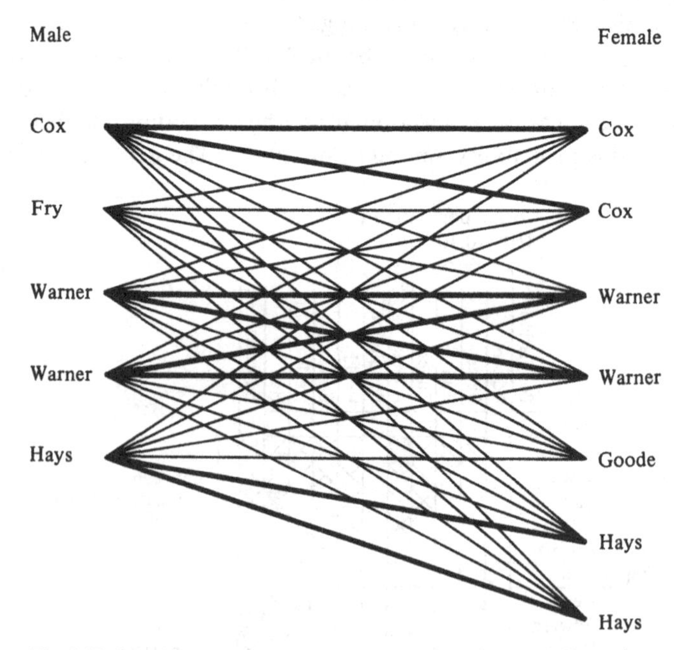

Fig. 2. Estimation of random inbreeding from isonymy. In this example there are 35 lines connecting a surname of a female with that of a male in some hypothetical population. In this case 8 (the heavy lines) connect individuals of the same surname. Under the assumption that 8/35 of all lines of descent are from ancestors in common, the random component of the inbreeding coefficient is one-quarter of this = 2/35 = 0.057.

One problem is that the degree of inbreeding is relative to some outbred condition. Some authors therefore adjust their inbreeding coefficients by isonymy within subpopulations to discount the degree of inbreeding that would be ascribed to the wider population. In fact the original study of Darwin (1875) and the study by Arner (1908) that followed it, discounted the general level of isonymy. V. Weiss (1973, 1974) presents a method for discounting the overall isonymy in his studies of inbreeding in a number of German towns. Surnames have a relatively short history, however, of about 30 generations; estimates of the genetic structure of populations can be cast in that time frame and need make no allowance for inbreeding prior to the adoption of surnames, of course. Alternatively if the extent of violation of the assumption of monophyletic origin of surnames can be estimated, results can be adjusted to that extent (Ellis and Starmer, 1978) or interpreted in the light of that knowledge.

Inbreeding coefficients from marital isonymy

According to the logic worked out by Crow and Mange (1965) the total inbreeding coefficient, F, can, under certain assumptions, be estimated to be the proportion of marital isonymy (symbolized by the letter I) divided by 4. Thus the relationship of F (inbreeding) to I (marital isonymy) is $F = I/4$. The random component of inbreeding, F_r, can be estimated as

$$F_r = \Sigma\, p_i q_i/4,$$

where p_i is the frequency of the ith surname in fathers and q_i is the frequency of the same surname as the maiden name of mothers. Since

$$F = F_r(1 - F_n) + F_n,$$

in which F_n is the non-random component of inbreeding, F_n can also be derived from

$$F_n = (I - \Sigma\, p_i q_i)/4(1 - \Sigma\, p_i q_i).$$

Dyke, James and Morrill (1983) have devised a way that could explain part of the non-random component of inbreeding. By selecting surname pairs taking into account the increased probability of mating with a spouse living at a shorter rather than a longer distance away and of the most likely ages, a higher frequency of isonymy is achieved than by using the Crow and Mange formula for the random component of inbreeding. Unfortunately Dyke *et al.* designate their value as an estimate of random inbreeding and use the symbol F_r for it. Their simulation also affects the estimate of the other component of inbreeding and of total inbreeding and again Dyke *et al.* use symbols that have been given different meanings. It might have been more helpful had they used their method of simulation to divide further the inbreeding coefficient into a random component (as before), a component due to distance effect, a component due to age effect, and a remainder due to inbreeding preference or avoidance within distance and age restraints. Likewise, Cavalli-Sforza and Bodmer (1971) suggest elimination of brother–sister pairs from estimates of random inbreeding. It is important not to be confused by the borrowing of the established terminology and symbols for different concepts.

Other slight variations in Crow and Mange's terminology and symbols have been proposed. Whether they are preferable depends on whether the lists of surnames used are considered to be the universe or a sample and on whether random inbreeding is estimated from males and females separately (as above) or from the sexes combined. Despite these variations and some minor flaws in some past publications, the basic logic has been reviewed and is sound (Crow, 1980). The assumptions are usually only partially met, however, and this must be recalled when interpreting results. The principal assumptions are that: (1) surnames are monophyletic and sharing a surname means sharing an ancestor from whom it was derived; (2) that at each degree of relationship matings along the male line are proportional to the number of kinds of lines; and (3) the two sexes are equally represented among migrants. When the extent of deviations from these assumptions is known, estimates of inbreeding from marital isonymy can be improved. But quantitative data on these points are hard to secure, and if one had such complete information there would be little reason to use estimates of inbreeding from isonymy because direct calculations could be made through path analysis.

The coefficient of relationships by isonymy

The random component of the inbreeding coefficient (F_r) when calculated from surnames is merely a statement concerning the average commonality of surnames (isonymy) between males and females in the population multiplied by a constant. The constant used is one quarter because this is the likelihood of a gene being shared by the homologous autosomal chromosomes of an offspring of first-degree relatives and the same fraction applies to other degrees of relationship by the logic adduced by Crow and Mange (1965). The likelihood of a gene being shared by first-degree relatives themselves is one in two. Therefore their coefficient of relationship by isonymy, *Ri* (and by extension of the Crow and Mange logic that of many other degrees of relationship also) is the proportion of isonymy multiplied by one half. As applied to the males and females of a population this is

$$Ri = \Sigma\, p_i q_i/2.$$

If one extends the logic and the assumption of the monophyly of the surnames to two populations this can be expressed as

$$Ri = \Sigma\, (Si_1 Si_2)/2\, \Sigma\, Si_1\, \Sigma\, Si_2,$$

in which Si_1 is the number of occurrences of the ith surname in a sample from population 1 and Si_2 is the number of occurrences of the same surname in a sample from population 2.

Morton *et al.* (1971) set forth this logic some time ago. They wrote that the kinship between populations may be estimated from isonymy between them and that the large number of surnames and absence of dominance give estimates from isonymy of high precision, but that the uncertain validity of the assumptions makes one hesitant to give the estimates disproportionate weight. The chief difficulty is that the more diverse the populations (geographically, ethnically, or in time) the less likely that a surname occurring in both had a common origin. Use of relative measures of relationship or inbreeding as suggested by Weiss (1973, 1974) and others (or at least interpretations based on relative rather than absolute values of surname commonality) will tend to avert this difficulty. In 1977 Weiss privately proposed that results of various methods be tested on the same body of data, but I did not follow up this sensible suggestion. One would expect that results by the different methods might differ considerably in absolute levels, but that they would correlate highly with each other.

Unlike the inbreeding coefficient by isonymy, the coefficient of relationship by isonymy is not divided into random and non-random components: it is a measure of the random component ($Ri = 2F_r$). If one wanted, the mean coefficient of relationship between pairs of specified kinds of individuals could be studied and the random and non-random components differentiated by exact analogy with the treatment of marital isonymy in Crow and Mange's study; there is little if any reason for such an analysis, however.

Another possibility with the coefficient of relationship by isonymy is to derive it separately for males × males, females × females, females × males and males × females. E. J. Clegg (unpublished) has done this to test whether one sex is more heterogeneous in geographic origin than the other. The four estimates between a single sex each of two populations have also been used to derive a measure of the variance of the between-population coefficient (Lasker *et al.*, 1979; Bailie, 1984). Other kinds of subgroups can also be separately analyzed (for instance, non-migrants × migrants, different time periods, occupational classes, educational levels, etc.) to assess long-term influences of these factors on the population structure.

More important, perhaps, the contribution of rare and common surnames can be separately measured (Lasker and Mascie-Taylor, 1983). The rare ones, which are more likely to be monophyletic, may yield interesting findings that would be obscured when rare surnames are included with common ones in a single coefficient. For instance, in the cited study, the rare surnames are much more sharply localized than the common ones.

Confidence limits of coefficients by isonymy

There is a problem in the study of marital isonymy: the variance is high and many differences shown by the method are therefore not statistically significant. Most isolates with very high proportions of inbreeding do not use European systems of surnames and they tend to be small and yield inadequate numbers for statistically significant results by marital isonymy methods. In less isolated communities the proportion of inbreeding averages 0.01 or less, giving proportions of marital isonymy of 0–4 per hundred marriages. In an example with 10 isonymous marriages in a sample of 300 – a sample as large as is usually possible for subpopulations – the confidence limits at $P = 0.05$ are roughly 6–15 in 300. That is, one could easily find 7 or 14 cases of isonymy rather than 10. With sample sizes and levels of inbreeding of this order, when the total inbreeding coefficient of one of the populations being compared is less than double that of the other, one cannot safely conclude that the difference is significant. Many published studies have smaller samples, lower levels of inbreeding, or make multiple comparisons and draw inferences that are not statistically significant: such results are merely descriptive and generalization from them is not justified.

The same problem does not apply to the same extent to the random component of the inbreeding coefficient derived from surnames nor to the coefficient of relationship by isonymy. For these the example given above provides 90000 comparisons of a male's surname with a female's and an additional 89400 of a male's with a male's or of a female's with a female's. Although these instances are not completely independent of each other, empirical evidence (e.g. Lasker, 1969; Bhalla and Bhatia, 1976; Lasker *et al.*, 1979; Kashyap, 1980) indicates very small standard errors. The non-random component of inbreeding has relatively the widest confidence intervals and least statistical significance since it is derived from marital isonymy and the total inbreeding coefficient. Experience suggests that only results of very large studies yield interesting information on the total inbreeding coefficient and the non-random component – especially when samples are subdivided into subsamples. On the other hand, samples of modest size can be divided into subperiods of time or into subgroups and still yield statistically significant differences in the coefficient of relationship by isonymy, random isonymy, and the random component of the inbreeding coefficient.

Although marital isonymy is less efficient than random isonymy in this respect, because sample sizes are often too small, sample sizes can be increased somewhat, with only slightly less genetic implication (and with the advantages pointed out by Wilson, 1981), if the isonymy of husband's surname with wife's mother's, and of husband's mother's surname with wife's mother's and father's are added to marital isonymy as usually defined (Lasker, 1968). Bhalla and Bhatia (1976) have shown that adding relationships of this kind also permits inbreeding estimates in societies with clan exogamy (a situation in which strict adherence to the rule would lead to an estimate of inbreeding of zero no matter what it actually was – as Reid (1973) has pointed out). Morton (1972, 1973) had already explored the use of isonymy with clan names; perhaps because the example from Micronesia may have seemed to others to pose ethnological problems, less attention has been paid to these publications than their theoretical interest warrants.

Some past studies have overinterpreted total inbreeding coefficients and coefficients of the non-random component of inbreeding. To avoid this one may put more stress on the coefficient of relationship by isonymy. The fine details of most studies in which inbreeding is partitioned by the method of Crow and Mange (1965), or modifications of it, can usually be adequately accounted for by chance factors influencing the number of isonymous marriages in subsamples: that is, the subsamples are too small for generalizations to be made from the results. Crow and Mange, however, did not subdivide the sample in their own original study; they were more aware of the need for a large sample size than some of their successors have been. Furthermore, Crow and Mange explicitly stated that the non-random component of inbreeding in their study of Hutterites is not statistically significantly different from zero. Some have cited the result but not this disclaimer; others have used the method and treated similar non-significant results as important.

Non-identical surnames

The last point to be made about methods here concerns non-identical surnames. Using the same sets of surname data as in studies of isonymy, the amount of information on non-identical surname pairings always far exceeds the amount of information on identical names. Morton (1972) used matrilineal clan names to estimate kinship in Micronesian populations. Using exchange matrices he examined the probability that a gamete in a clan *j* had come from clan *i*. Subsequent studies of surname transitions have not cited this. Devor (1983) has applied a different method: the transition of a wife's surname to that of her husband. It is like a migration matrix that considers only the movement of females. This surname-pair-matrix method has also been used by Koertvelyessy *et al.* (1984). When it is applied to marriages, the association of the bride's maiden name with the surname of the bridegroom can be thought of as the transition involved in her adoption of his surname. The associations between families can then be estimated as the 95 % or 50 % passage time (in generations) as has been done for genes from migration data by Hiorns *et al.* (1969), Coleman (1980*a*), and Lalouel and Langaney (1976). The spread of certain surnames throughout the country from a place of origin seems to be slower than would be predicted from a migration model, however, probably because offspring of migrants are more likely to move than the offspring of sedentes, and when sons of migrants do move they move back to the place of their parents' origin more often than would be expected at random. Such non-random behaviour among families would leave more genes (and surnames) at the place of origin than if migration in each generation were random with respect to the prior generations' migrations. In fact surnames do tend to be more frequent near their points of origin than one would expect from the amount of migration (Kaplan and Lasker, 1983; Lasker and Kaplan, 1983). These studies were limited to surnames from place names, however, and the full extent of the phenomenon may have been somewhat obscured by including surnames from cities with those from villages and by the inclusion of some surnames that derived from more than one place of origin.

The use of the full matrix of maiden names of brides by surnames of bridegrooms presents some special problems. Devor (1983) pointed out that the size of the matrix of surname-pair combinations is very much larger than the number of individual pairs (marriages). For instance, Yasuda (1983) cited studies which estimate that there are about 120000 different Japanese surnames in a population of about a hundred million. The several tens of millions of marriages would therefore be distributed among over ten thousand million possible surname combinations and over 99% of the possible pairs of two surnames would never occur in any marriage. A similar situation would occur in almost all other national and smaller populations.

To attempt to deal with this, Devor grouped the surnames into five categories according to relative commonness. Unfortunately, the finding of approximately random mate exchanges among such categories (e.g. Devor, 1983; Devor *et al.*, 1984) has little bearing on the question of the influence of family name on mate choice. Each of the five categories of surnames is so diverse in its contained names that, in the great majority of instances, two marriages in the same cell in the matrix will have a different pair of surnames. In cases where the mate exchanges among categories of this kind are not at random, the associations may well rest on features associated with surname frequency. For instance immigrants are likely to have rare or unique surnames and are more likely than the rest of the population to marry other immigrants.

An alternative strategy would be to group surnames at random and to do it repeatedly to build up sample sizes. However, this would not avoid the basic problem of the heterogeneity with respect to surnames of each matrix category. One could avoid this heterogeneity by searching the list of marriages for all instances where the same pair of surnames occurs repeatedly. The number of such repetitions could then be compared with the number expected at random in a population with the same surname frequencies. An excess of observed over expected repetitions could result either from marriage preferences closely associated with surnames (such as a pair of siblings marrying a pair of siblings) or social or geographic factors (correlations in the distribution of surname frequencies with occupation or with location in the area studied).

The rate of repeated pairs of surnames of couples measures something akin to the Wahlund effect of subdividing a population; in Paracho, Mexico, B. A. Kaplan and I find a rate of 0.00086 (29% more than expected at random). This is far less than the extent of isonymy (0.015), but has a lower proportional standard error of estimate.

This raises the question of regional and national studies of surname distributions. This wider scope permits application of higher-order methods such as the correlation of coefficients of relationships by isonymy with the distances between the places related (Lasker, 1978*a*; Souden and Lasker, 1978; Küchemann *et al.*, 1979; Raspe and Lasker, 1980; Lasker and Mascie-Taylor, 1983). Direct mapping of surname frequencies and the analysis of such maps by regressions against latitude and longitude will be covered in Chapter 11 and the Appendix.

5 *Isolates and inbreeding*

Geneticists have long been interested in human inbreeding because of the known deleterious effects of close inbreeding (such as repeated sibling matings in domestic cattle). The effect of the lower levels of inbreeding that occur in human populations is less clear. Rare recessive genetic traits are found principally in the offspring of consanguineous marriages: for instance most cases of fructosuria occur in offspring of related parents (M. Lasker, 1941). Because such conditions are very rare, however, they have little importance. The significant question is whether inbreeding between first and second cousins (the usual levels in human populations) increases the risk of morbidity and mortality from common disorders. There is some controversy about it but the best large studies seem to show a small but significant effect (Schull and Neel, 1965).

The interest of biological anthropologists in inbreeding has usually been in the basic questions of the kinds and amounts of inbreeding that occur in the isolated populations that traditionally interest them. This has led to studies of small societies, such as that by Simmons *et al.* (1962, 1964) of the Aborigines of several islands in the Gulf of Carpentaria, Australia, that had been cut off so completely from each other and from mainland Australia that they had quite divergent gene frequencies. In South America, Ward and Neel (1970) found that seven villages of Makiritare Amerindians were virtually as diverse in frequencies of blood group genes as are the tribes of Central and South America from each other. These and other such groups do not lend themselves to surname analysis, however. Just because they are so isolated they have had no need for surnames.

There are some islands, though, where populations of European cultural traditions have been isolated, and surname models of inbreeding should be applicable to these. One of them, Pitcairn Island in the Pacific, has a population that was founded in 1790 by six of the mutineers of the British ship, the *Bounty*, and 13 Tahitian women (Refshauge and Walsh, 1981). Since then many people have left the island (notably a large exodus to Norfolk Island) but few outsiders have come to Pitcairn to live. Interestingly, the isolation of Pitcairn has increased in recent years; there is no landing strip and there are now fewer visiting ships than formerly. Some six or seven generations have passed since the colonization and the present population of 50 individuals has only four surnames – those of three of the mutineers and that of a whaler who settled there. Half of the surnames of the original male settlers have been lost. The phenomenon of surname loss through chance extinction of male lines is almost inevitable in small populations. Galton (1889) dealt with statistical aspects and the theory was developed by Lotka (1931) and others and refined by Yasuda *et al.* (1974) in a study of the population of the Parma Valley in Italy. When a surname is confined to a few individuals there is a considerable possibility in any one generation of none of the males having a male heir. Once a surname is held by a fair number of individuals, however, the likelihood of its being totally lost becomes very small indeed.

The classic study of a population by isonymy is that of the Hutterites by Crow and Mange (1965). There were 15 surnames among them of which 11 appeared in the subpopulation (*leut*) which was studied. The names are of German origin and were brought to America in 1879 by the already somewhat inbred founding members. By the method described earlier, Crow and Mange calculated from the fact that 87 of 466 marriages were isonymous that the inbreeding coefficient was about 0.0495. This compares with an estimate from the pedigrees of 0.0226. The random component of the inbreeding coefficient by isonymy was 0.0455 and the non-random component, 0.0052. The pedigrees extend for five generations so measure inbreeding through fourth cousins: this accounts for 46% of the inbreeding estimated by isonymy. The remaining 54% is ascribed by Crow and Mange to relationships more remote than fourth cousins, that is, to a period preceding the Hutterites' migration to America.

Yasuda and Morton (1967) extended the study and exploited one of the special advantages of surname analyses in genetic research – that the method can be applied to historical as well as contemporary data – to distinguish between the founder effect and subsequent inbreeding among the Hutterites. When the Hutterites had sought refuge from persecution by migration to Russia in 1762 there had been 15 surnames among them, the frequencies of which yield an estimate of the random component of the inbreeding coefficient at that time of 0.117. Among those who migrated to America in 1879 or 1880 the equivalent coefficient was 0.33 and by the married generation in 1960, one of the 15 surnames of the group studied was no longer represented at all and, according to the data analyzed by Yasuda and Morton, the random component of the inbreeding coefficient calculated from isonymy had risen to 0.040. Thus about

42% of the loss of 'allelic' variability through genetic drift was due to the founder effect in the original migration to Russia. Furthermore, during the 80 years preceding the study – a period when inbreeding was decreasing in most populations of the world – the Hutterites were becoming more inbred despite a phenomenally high rate of natural increase in the size of the population.

Another religious isolate in the United States with a somewhat similar population structure is the Amish. Hurd (1983) found nine surnames in one group of Amish. This is similar to the 11 surnames in one group of Hutterites and suggests that the two isolates may have had similar rates of inbreeding. The reported level of isonymy and inbreeding coefficient in the Amish are, however, somewhat higher than those of the Hutterites. The fissioning and migration of some colonies of Amish from Pennsylvania to Indiana and elsewhere have resulted in certain of the characteristic Amish surnames being more common in one of the areas where they now live and different surnames being more common in a different area. Hurd (1983) studied three endogamous Amish churches, each composed of two or three districts. Of the only nine different surnames in all, three were held by one woman each and will disappear in the next generation. Hurd compared the coefficient of relationship by isonymy between districts within the same church (mean $Ri = 0.1653$) with that between churches (mean $Ri = 0.1465$) and found no significant difference despite the fact that there was a significant difference in the index of relatedness calculated from the pedigrees by Wright's method (mean = 0.0725 and 0.0542, respectively). The correlation between results by the two methods was low either because of the high variance of Ri or because of the difference in what is being measured by the two methods. The fit of the surname frequency distribution by the method suggested by Fox and Lasker (1983) or by that of Zei *et al.* (1983*b*) would also be very poor because the two most common surnames occur in very nearly equal numbers (154 and 149) which no other surname approaches (the next two yielding 30 and 26 cases respectively). Thus some 95% of the random isonymy would be from the two commonest surnames, and small random variations in their distribution in districts would have large effects on coefficients based on isonymy. So this is one of those instances mentioned in the chapter on methods where interpretation of findings should be limited to the summation over a number of studies and where any effort to show details of the population structure by isonymy would be of limited reliability.

There are many religious isolates in India, both among endogamous Hindu castes and in tribal isolates. The various naming practices are different from those in Europe, however, and the method of isonymy can only be applied to some and even then requires thoughtful decisions as to what constitutes a surname. For instance, male Sikhs generally bear the name Singh and females the name Kaur, so if one considers these as two surnames there would be no marital isonymy (except in the rare cases outside India where official records show the surname of the daughter of a Singh as Singh). If one considers Kaur and Singh as variants of a single surname, however, marital isonymy would be extremely common among Sikhs. If one considers Kaur and Singh not to be surnames, it might be difficult to determine the surnames from available records.

Some Indian caste names are used like surnames and endogamous marriages (the usual kind) all appear isonymous. However, mobility in the caste system is as old as the system and such mobility is usually carried out or accompanied by a change of surname.

Despite these problems, Bhalla and Bhatia (1976) successfully modified the isonymy method to take account of clan exogamy. In one North Indian group, the Bhatia, the classificatory cross-cousin marriage system, in which cousins of whatever grade traced through an ancestral brother–sister pair are preferred partners in marriage, led to 63 of 345 couples showing isonymy between husband's father and wife's mother, but there was no marital isonymy. Almost all the inbreeding shown by isonymy in this case can be accounted for by the random component, as would be expected in a group with almost complete endogamy marrying at random among themselves. Bhatia and Wilson (1981) and Wilson (1981) have further developed the logic of a generalized isonymy which substitutes isonymy of other relatives for that of husbands with wives – an approach previously tried by Lasker (1968, 1969). Wilson showed that use of surnames of the four parents of a married couple allows estimates of inbreeding even when clan endogamy is tabu, and the method also aids finding other elements of the breeding structure such as preference for cross-cousin marriage. It should be noted that Yasuda and Morton (1967) reported that Cotterman had given them a personal communication in which the isonymy method was generalized to cases where the proportion of isonymy is some ratio of the inbreeding coefficient other than $I = 4F$.

Kashyap (1980) applied inbreeding coefficients by isonymy to names used by the Ahmadiyyas of the Kashmir Valley. The non-random component is high, but the numbers on which it is based are not reported. In the most recent generation, for which the pedigree information is most nearly complete, the random component (0.0269) is very similar to the inbreeding coefficient calculated from pedigrees (0.0287). Kashyap and Tiwari (1980) report values for the coefficient of relationship by isonymy between Ahmadiyyas villages. If these values are divided by 2 to make them comparable to the previously published inbreeding coefficients (an adjustment made necessary by the different constants in the two formulas), the values among villages tend to be slightly lower than but very similar to the random inbreeding coefficients calculated from the isonymy data for the group as a whole, and are hence also very similar to the inbreeding estimate for the most recent generation as calculated from pedigrees.

Muslim populations also pose special difficulties because they do not use surnames in the European sense and because, in the Near East, there is a preference for marriage with patrilateral (male line) kin and the ratios assumed by Crow and Mange's method therefore do not hold. Lewitter, Hurwich and Nubani (1983) applied calculations of isonymy to first names in an Israeli Arab community. The community is divided into six *hamulas* or clans and people are known by their own, their father's, their father's father's given names and so on back to the clan founder. Since given names are not often unique, inbreeding could not be traced through isonymy, it was said. The authors did not point out, however, that there are six clan names and that these indicate direct male ancestry and are the equivalent of surnames. Using them as such and the formula

$$F_r = \sum Si(Si - 1)/4n(n - 1),$$

in which Si is the number of children with the ith clan name and $n = \sum Si$, I calculate a random inbreeding coefficient for 899 children born in 1950–74 of $F_r = 0.061$, which is comparable to $F = 0.054$ calculated by Lewitter *et al.* from pedigrees. However, the patrilineal great-grandfather of the founders of the six clans, Abu Ghosh, gave his name to the village. If one considers Abu Ghosh to be a surname, it is shared by all the members of the community and their random inbreeding coefficient would be calculated by the formula given above as the unreasonably high figure of 0.25. Thus one may agree with the authors that without the help of genealogical information the percentage of common ancestry in such a society cannot be calculated from a knowledge of the people's names.

Ethnic isolates exist in many societies. In the United States some societies of mixed Amerindian descent have a separate existence and considerable inbreeding. In one, the Melungeons of Tennessee, Pollitzer and Brown (1969) confirmed by isonymy the high level of inbreeding. In some such cases the isolated environment and ethnic distinction reinforce each other and both contribute to a high rate of endogamous marriage and consequent inbreeding.

High mountain valleys are not cut off from the lower slopes of their drainage systems the way mid-ocean islands are separated from the continents, and mountain villages usually have no religious proscriptions against marriage with others. Nevertheless, the difficulties of travel in the mountains, the frequent national and linguistic barriers along mountain crests, and the sharp ecological distinction between herding in the uplands and agriculture below, have imparted a high level of isolation to the human populations of the valleys in the Alps. The structure of the human populations of several of these valleys has been studied by analysis of their surnames.

Hussels (1969) collected genealogical material from Saas, a Swiss valley, and compared the extent of inbreeding implied by isonymy with the amount known to have taken place. In this case, isonymy overstates the inbreeding because of the virilocal marriage system (the women usually move to the husband's place of residence and the men tend to stay where they inherit land, cattle and grazing rights). Thus a few surnames become and stay dominant locally and isonymy tends to be higher than would otherwise correspond with the inbreeding level.

Friedl and Ellis (1974) and Ellis and Friedl (1976) conducted a similar test in another German-speaking Swiss valley, Kippel, and Ellis and Starmer (1978) tested a third, Törbel. These studies together permit measurement of the differences between estimates of inbreeding from isonymy and from pedigrees and an evaluation of the causes of the differences. The estimates of inbreeding from isonymy are higher than those from pedigrees in all of these cases. The difference stems in part from polyphyletic surnames (surnames founded independently more than once) and in part from the virilocal marriage system as a result of which men are less mobile than women. If the system were strictly followed it would limit the surnames of a community to the same few that had always been there (less any that disappeared through families without male heirs). This would hold even if numerous females from elsewhere regularly married in. In these careful studies of Swiss communities only a little of the difference between inbreeding rates calculated from isonymy and those calculated from pedigrees can be accounted for by the incompleteness of the pedigree records; distant consanguineous matings that may have been missed (especially if the ancestor in common was not a resident of the community) are only one of the causes of instances of isonymy that are missed in the pedigrees.

Six communities in the western Alps in Italy also provide surname information used in studies of population structure in mountains (Fig. 3). Crawford (1980) reported on marital migration and isonymy in three communities in an Italian Alpine valley, Val Maira. Elva is on a side

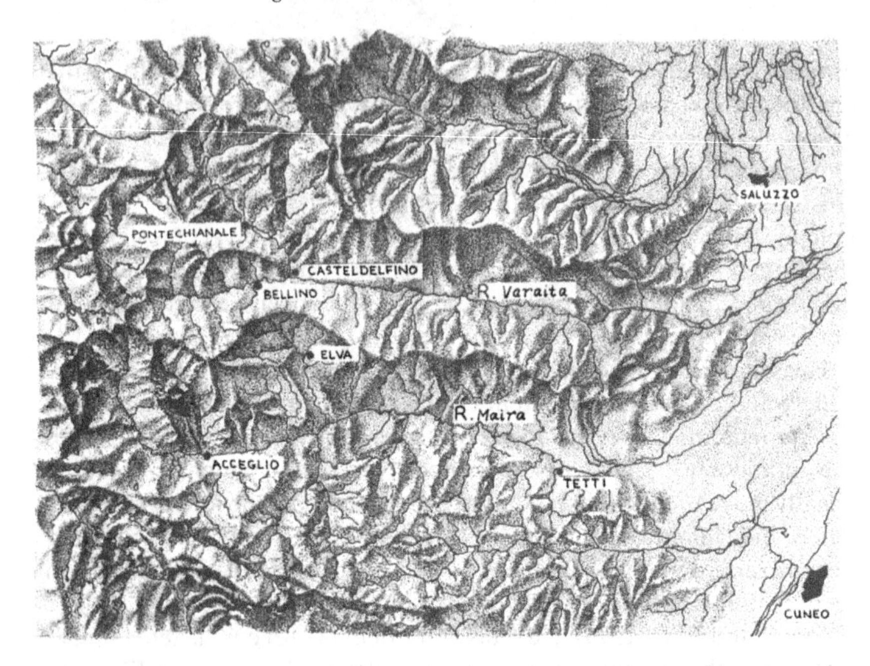

Fig. 3. Alpine valleys in western Italy, showing communities in which studies of isonymy were carried out.

valley near a ridge from Mt Pelva and the only access is via steep and treacherous mountain roads. Acceglio, at the head of the valley, is on a somewhat better road that continues, a few kilometres further on, to France. And Tetti is near the opening of the valley into the broad agricultural plains of Piedmont. Endogamous marriages and isonymy in each of four 20-year periods were highest in Elva, next highest in Acceglio and lowest in Tetti, and the proportions decreased, with few exceptions, in each time period. The proportion of isonymy (presumably in per cent) is reported to have been 21.50 in the first period and 12.60 in the last at Elva; 15.12 in the first period and 4.13 in the last at Acceglio; and 9.00 in the first period and 2.10 in the last at Tetti.

We conducted similar studies in three communes in the next valley, Val Varaita, only 14 kilometres to the north by path over the mountain, but some two hours by car on the paved road down one valley from Acceglio and up the next, or on a dirt road over the mountain from Elva. The Varaita River arises among steep mountains that form the frontier between France and Italy. Along the southern fork are the nine hamlets, *frazioni*, of Bellino with a total population in 1961 of 546. On the northern fork, Pontechianale, now developed for skiing, had ten *frazioni* with 400 inhabitants. The confluence of the forks of the Varaita was guarded by a mediaeval castle and the town of Casteldelfino with 12 *frazioni* and 650 inhabitants in all. Travel in the valley is along the roads which closely follow the river and its tributaries, since the narrow valleys are hemmed in by 3000–4000 metre high mountains on the north, west and south and a ridge also separates Bellino from Pontechianale except by way of Casteldelfino.

In Bellino, at the head of the valley, the more isolated of the two communities, use of a local dialect akin to Langue d'Oc distinguishes the people from those of lower parts of the valley. In Bellino in 1951–70, 6.65% of marriages were isonymous and a century earlier, 13.82%; in Casteldelfino, the next community down the valley, the corresponding values were 5.56% and 11.59% (Lasker *et al.*, 1972). The rates of endogamous marriages showed the relative isolation of the earlier over the later period and of Bellino over Casteldelfino to be even more pronounced. Marital isonymy was at about the same level as isonymy between other affinal relatives (husband vs. wife's mother, husband's mother vs. wife and wife's mother, and husband's and wife's fathers vs. mothers). Observed isonymy was generally somewhat greater than random isonymy so there was a positive non-random component in some but not all time periods in both communities. As one would expect with increasing exogamy, most of the accumulated inbreeding is explained by the past (random) component rather than the current choice of mates at the times covered in the calculations. Of the 64 marriages between persons born in the two contiguous communes, one was isonymous (close to the expected rate of 1.57% based on the frequencies of surnames shared between them).

We subsequently studied the relationship between the *frazioni* of the communes and between the communes as a whole (Kaplan *et al.*, 1978). The coefficient of relationship by isonymy between *frazioni* of the same commune tended to be high. In 97 of 150 cases, including all those within Bellino, *Ri* was greater than 0.01; it was also above 0.01 between adjacent *frazioni* of different communes. Values of *Ri* between adjacent communes with centres about 6 kilometres apart by road are 0.0071 (Bellino–Casteldelfino) and 0.0083 (Casteldelfino–Pontechianale), whereas between the two communes separated by about 6 kilometres of mountain but 12 kilometres of road, *Ri* was 0.0038 (Bellino–Pontechianale). These rates are based on equating French forms of the surnames with Italian ones, because the people of Bellino (Blins, in local dialect) use the former and those of Casteldelfino and Pontechianale the latter.

Bourgoin and Vu Tien Khang (1978) studied the genetic demography of four villages in the Pyrenees in France by analysing the parish records from 1740 to 1792 and the civil registers thereafter until 1973. The migration matrices demonstrated that in three of the villages the majority of genes present ten generations ago would have been displaced by ones introduced by immigrants during the period studied. The fourth village, Le Clat, however, would still have retained the majority of its original genes. This was reflected also in three types of measure of inbreeding: (1) close inbreeding recorded in ecclesiastical dispensations for marriages of relatives; (2) total inbreeding from pedigrees; and (3) inbreeding from marital isonymy. The rates from indulgences and pedigrees are much below the levels calculated from marital isonymy. However, among 81–180 marriages per village per period, there are only 1–21 (median, 6) isonymous marriages. Thus in samples of this size proportions of marital isonymy between about 2 and 8 % would be expected at random: only the highest values are out of this range. The greater isolation of Le Clat than of the other three villages shown by the lower proportion of exogamous marriages in the former is confirmed by the isonymy data: in each of four periods the proportion of isonymy is higher in Le Clat than in any other village in any period. One cannot expect to be able to answer more detailed questions, such as those about changes within individual villages with time, by proportions of marital isonymy in these ranges in villages of these sizes.

Unfortunately the more precisely ascertainable random component of inbreeding by isonymy and coefficients of relationship by isonymy between the villages were not calculated. Only the unusually high inbreeding rate at Le Clat relative to the other three villages aligns the isonymy results with those from the pedigrees. The isonymy data appear to correspond better with the migration data than do those from pedigrees and indulgences, however. In villages of these sizes (between 200 and 600 persons per village until a decline in the present century to the 50–150 range) random endogamous mating is an important component of the inbreeding coefficient and this is seen in proportions of marital isonymy and would be even better expressed in the proportion of random isonymy. There is no gainsaying the fact, however, that inbreeding levels from pedigrees and indulgences as low as 10 % or so of those from marital isonymy, as in some of these cases, suggest that the latter is inflated by surnames being polyphyletic, by patrilateral kin marriages being more frequent than the assumed ratios, by sex differences in migration patterns, or by other departures from the assumptions necessary to equate inbreeding calculated from marital isonymy with the rate from complete pedigrees.

Rogers (1984) has compared estimates of inbreeding from isonymy

Table 1. *Inbreeding coefficients measured by isonymy and pedigree methods*

Population	Source	Pedigree *F*	Isonymy *F*	Ratio (%)
Hutterites	Crow and Mange (1965)	0.0226	0.0495	46[a]
Saas	Hussels (1969)	0.00536	0.02049	26[b]
Swiss	Morton and Hussels (1970)	0.00062	0.00625	10[b,c]
Kippel	Ellis and Friedl (1976)	0.009	0.032	28[a,c]
Törbel	Ellis and Starmer (1978)	0.00496	0.02510	20[a,c]
Ahmadiyyas	Kashyap (1980)	0.0287	0.0973	29[a]
Orkney	Roberts and Roberts (1983)	0.000195	0.00146	13[a,b,d]
St. Bart	James (1983)	0.009	0.0271	33
Amish	Hurd (1983)	0.0756	0.0249	39[d]
Vinalhaven	Sorg (1983)	0.0003	0.0012	25[b,d]
Mennonites 1800	Rogers (1984)	0.0162	0.0159	102[e]
Mennonites 1980	Rogers (1984)	0.0027	0.0036	75

[a]Explained by remote consanguinity included by isonymy but not by pedigree method.
[b]Explained by polyphyletic origin of surnames.
[c]Explained by patrilineal inheritance of land and virilocal marriage pattern.
[d]Sampling variance discussed.
[e]Based on pairs of parents. If all individuals were used for calculation of *F* from pedigrees, the ratio would be lowered to 39%.
In part after Rogers (1984).

with those from pedigree analyses. Her list can be further extended by recalculations from other published data (Table 1). In the various previous studies the values calculated from pedigrees ranged from 10 to 46% of those from isonymy. In her own studies of Alexanderwohl Mennonites, however, after adjustment for cumulative inbreeding, Rogers found values from pedigrees were about 35–39% of those from isonymy based on all individuals or on all parents, but 75 and 102% (for different periods) when isonymy is calculated from the surnames of the children, i.e. from the same population as that from which the pedigree inbreeding is calculated. Thus in this case isonymy and pedigree estimates seem not to be greatly different.

6 *Island versus distance models: the Far East and Oceania*

There are two types of model of the geography of population structure on which theoretical constructions are based: the island model and isolation by distance. Both have been used by population geneticists to model inbreeding in human populations. In actuality human populations tend to fit one or the other type of model better but almost always have elements of both. The island model assumes that individuals aggregate in tight clusters with essentially random mating within the cluster ('island') and with greater genetic differences from all other such clusters than among themselves. In this model the individuals share in the gene pool of their own 'island' in equal degree. This is a model which fits human isolates reasonably well and is implicit in most of the studies cited in the previous chapter.

The alternative model, of isolation by distance, assumes that space (conceived as a two-dimensional surface in the case of terrestrial organisms) is more or less evenly peopled by individuals whose ancestry and genetic constitution vary according to their position on the surface. The difficulty with this model when applied to humans is the obvious clustering of people in cities, towns and villages. In addition, genes and surnames are discrete entities with all-or-nothing occurrence. Since any point in space can have only one individual at it, the probabilities of occurrences at each point have to be calculated from other occurrences not at that point but considered relevant because of their spatial relationship to it. Therefore when an isolation-by-distance model is used, space has to be grouped into segments for study. In human studies the gene frequency or surname frequency of a segment is assigned to some point within it (such as a town centre) which is thought to be the centre of the population of the area. A model based on such data is reasonably appropriate where the areas of the segments are small relative to the area of the whole study and where the habitat and the mode of its human exploitation dictate relatively homogeneous settlement patterns, as is the case with farming. The model is therefore relatively compatible with regional studies of peasant societies in open country without severe water or mountain barriers.

The problem with the use of an island model, even for isolated populations, is illustrated by a study of the small French West Indies island of St. Barthelemy by James (1983). Twelve of the 97 marriages during 1951–60 were isonymous, yielding an established inbreeding coefficient of 0.314. However, when computer simulations of marriages were run, it was necessary to introduce a factor for geographic distance between residences of potential spouses. This and allowance for the ages of the potential partners increases the estimate of random isonymy so that the estimate of non-random isonymy becomes negative rather than positive. This suggests that the non-random component of the inbreeding coefficient in this population (when calculated without these allowances) results in part from the propinquity of mates rather than from any preference for marriage with kin or for marriage with persons of the same surname. Although this point was not further explored by James, it illustrates the kind of issue that can be approached by the application of isonymy.

In studies of basically agricultural societies in the Far East the application of surname models spans the whole range of variation of human societies with respect to variety of surnames and time since their introduction. Chinese surnames were in use in the Han Dynasty (at about the time of Christ) and are therefore the most ancient in general use anywhere in the world. Chen and Cavalli-Sforza (1983) refer to a list of Taiwan surnames, collected by S. H. Chen and M. H. Fried, that contains all 1195 common Taiwan surnames. Since the present population of Taiwan includes large numbers of immigrants from all parts of China, this list must contain all or virtually all the surnames regarded as Han Chinese on the mainland. Surnames of some ethnic minorities may not be so completely included, however. The analysis by Chen and Cavalli-Sforza is of all surnames, but the population is so dominated by descendants of the mostly eighteenth-century immigrants from Fukien Province and the few surnames they bear that results of this study would not have altered appreciably had all other surnames been excluded. The possibility of a separate analysis of these other classes of surnames was not exploited. Although the original data were given by smaller administrative units (which might be expected to reflect the effects of recent local marital migration), the authors limited their study to larger units, the *hsien*, and eliminated the cities from consideration. From a matrix of inter-*hsien* random isonymy there is virtually none of the negative correlation of log isonymy with distance that has been claimed for genetic similarities at various distances (Malécot, 1948). However, higher-order analysis by branching tree diagrams and principal components is interpreted to reflect something of the original pattern of settlement from mainland China and of subsequent fission of the colonizing populations.

Studies of inbreeding in Chinese populations based on proportions of marital isonymy would be meaningless because of certain aspects of the naming and mating systems in China. Marriage between persons of the same surname – whether related to each other or not – is tabu, hence marital isonymy must be extremely rare and bears no relationship to overall inbreeding rates. At the same time clans often dominate villages so that virtually all the residents of a village (other than females who have married in and whose offspring will not bear their surname) share one or a few surnames. Depending on how it is calculated, random isonymy may be high and it will bear little if any relationship to actual inbreeding levels.

Japanese surnames stand at the opposite extreme from Chinese ones in respect of two characteristics of importance for genetic models. Whereas there are only a few Chinese surnames, Japanese surnames are abundant. Whereas Chinese surnames are ancient, most Japanese surnames are very modern. According to Yasuda's (1983) survey of Japanese publications, there were about 30000 Japanese surnames before 1875 when a cabinet decree made surnames obligatory for everyone, and the number of different surnames had increased to 120000 by 1981. All but six of the common Chinese surnames consist of a single Chinese character (in fact other multiple-character surnames in China are assumed to be of foreign origin); on the other hand, Japanese surnames usually consist of combinations of two or three Chinese characters, so the number of permutations possible is very large. Nevertheless, the local Buddhist priests who gave surnames to their parishioners about a hundred years ago often gave the same lucky set of characters to several families, so different lines of a polyphyletic surname sometimes arose in the same place. Polyphyly has therefore affected within-population as well as between-population isonymy from the outset of surname use in Japan.

On the basis of various lists, Yasuda and Saitou (1984) estimated that random isonymy for Japan as a whole is 0.0022 (that is, the coefficient of relationship is 0.0011). In local studies the frequency of isonymy is much higher, however, and Yasuda and Furusho (1971*a*,*b*) estimated the inbreeding coefficients as 0.0011 and 0.0023 in two small cities and an average of 0.0044 in a group of farming villages. Yasuda and Saitou (1984) state that random isonymy in Japan does not necessarily reflect genetic inbreeding. The reduction in isonymy in recent times seen in some Japanese studies is said to be associated with non-random, not random, isonymy. Of course the way isonymy is partitioned means that non-random isonymy has a relationship to inbreeding in the current generations whereas random isonymy relates to prior generations as well. Other Japanese studies are summarized by Yasuda (1983).

Katayama, Toyomasu and Matsumoto (1978) studied the surnames of the population of Kamishima, a Japanese island with a 91.9% rate of endogamy. The proportion of marital isonymy, 0.2283, was not statistically significantly different from random isonymy (0.2154) or from isonymy of wife with husband's mother (0.2400), husband with wife's mother (0.2529) or husband's mother with wife's mother (0.1944). At least one of the assumptions of inbreeding calculations by Crow and Mange's method (that there be little if any preference for or against marriage with a person of the same surname) is apparently well met. However, the proportions of isonymy and of random isonymy were much higher in Kamishima than in other populations Katayama *et al.* had studied (0.02954 and 0.05635 respectively for Hatomajima and 0.02752 and 0.03288 for Kuroshima). The authors believe that there was little polyphyleticism of surnames on Kamishima because, when surnames were ascribed, each surname was given to sets of relatives. This would mean that the degree of interrelationship implied by the high inbreeding coefficient calculated from marital isonymy (0.057) is largely due to remote consanguinity (modern marriages of persons related to each other through common ancestors over 100 years ago). In fact, the authors state that there was a low frequency of first cousin marriages but much marriage of more remote relatives such as second cousins, second cousins once removed and third cousins and of multifold relations such as double second cousins. The high level of isonymy on this island is consistent with the evidence of random genetic drift which the authors report in the frequencies of genetic markers there.

Katayama and Toyomasu (1979) also compared evidence concerning surname commonality, geographic distance and gene frequency differences of Kamishima, two populations on a somewhat larger island and a fourth population on a major island. The branchings of tree diagrams of these relationships are identical for divergence of net surname differences and gene differences, and the extents of the differences correlate positively ($r = 0.69$). Assuming that the surname commonality is due to past migration, the r^2 value of 0.47 implies that 47% of the genetic variation can be ascribed to migration and the rest to random genetic drift. Since the genetic differences bore no significant relationship to geographic distances between the populations, an island model of population structure fits these populations better than an isolation-by-distance model. There is some relationship between surname commonalities and interpopulation distances, however, as noted in other parts of the world.

Yasuda and Morton (1967) compared the random component of the inbreeding coefficient of various Far Eastern and other ethnic groups in Hawaii. The Chinese plus Korean surnames gave a figure of $F_r = 0.0066$; the Japanese, 0.00051; the Caucasians, 0.00032; the Hawaiians, 0.00029; and the Filipinos, 0.00028. As already noted, the Chinese have very few surnames and consequently a proportion of random isonymy over ten times that of any of the other groups. The Japanese have a proportion similar to those reported for Japan as a whole. And the Caucasians have a proportion somewhat higher than is typical of the mainland United States or of Great Britain as a whole.

Two districts in rural central Tasmania gave results that imply they are relatively outbred (Kosten *et al.*, 1983). Isonymy was low and random isonymy even lower – especially between the districts – so if reliance is placed chiefly on the random component, this population will be seen as much more outbred than its geographic position 'at the ends of the earth' might seem to make likely. The total inbreeding coefficient and its non-random component in these two districts are calculated from 16 cases of isonymy among 1179 marriages; even if these are not partitioned between the two districts and various periods of time, the span between the confidence limits of the rate would be large and the finding of appreciably more total than random isonymy need not lead one greatly to modify the interpretation from the latter of an outbred population structure.

7 *The Americas and continental Europe*

Modern mainstream United States populations have been little investigated by surname analyses. A sample of names in a suburban Detroit telephone directory yields a low proportion of random isonymy, as one would expect from the heterogeneous ethnic origins of such populations. However, they are not true breeding populations and actual matings may be considerably more inbred. The diverse origins of Americans and consequent large numbers of different surnames – said to be over a million – imply low levels of isonymy. Because of considerable polyphyly of surnames and the high rate of name-changing that has characterized immigrants to the United States, the random component of inbreeding is probably even less than implied by random isonymy. However, in the early stages of immigration of each new group, the group seeks to keep its identity and members tend to marry among themselves, so the inbreeding rates may be similar to those of the country of origin. In fact, if the immigrant group is small, inbreeding may be more intense than in the place of origin. This would be reflected in the non-random component of inbreeding in the first few generations. Except for a few groups with very strong religious sanctions against intermarriage, such immigrant enclaves in the United States dissolve after one, two or three generations in the melting pot of urban America.

The classic study of Arner (1908) in the United States has already been cited. Arner's data and Crow and Mange's (1965) method give a total inbreeding coefficient of about 0.0052 and a random component of about 0.00019 for New York in colonial times and a total inbreeding coefficient of 0.0021 for one county in Ohio during the nineteenth century.

Swedlund (1971) applied the method of isonymy to early historic data on the towns of the Connecticut River Valley in Massachusetts Colony and he has discussed the values and shortcomings of the method (Swedlund, 1975). In a recent repeat study (Swedlund and Boyce, 1983) there were 140 isonymous pairs among 9473 marriages. The assumption that the surnames were monophyletic was tested and reasonably well met in this study, but the assumption that males and females migrate in equal proportions did not hold true. There were more female migrants than male, but the males generally migrated longer distances (as seen also in some other studies: e.g. Lasker *et al.*, 1972). As in most studies of subdivided populations, Swedlund and Boyce found random isonymy between populations (Ri = 0.00138) to be considerably less, on average, than that within populations (Ri = 0.00450). A few surnames were associated with wealthy families; among these elite families isonymy was high (0.909 versus random expectation of 0.0048) and notably so in town-exogamous marriages.

The only other study of colonial North American surnames is an unpublished one by A. T. Steegmann and G. W. Lasker of the members of several companies of militia in the Colony of New York. There was a very low level of the mean coefficient of relationship by isonymy between companies raised in different counties. This reflects the wide diversity of origins of the colonists, who came from England, Ireland, Germany and other countries. On the other hand there was a high average coefficient of relationship by isonymy within the separate companies, largely because brothers and other close relatives were often enlisted together.

There are only a few studies of modern North American populations other than those of isolates, which have already been described, and those of Spanish-surname enclaves, which will be discussed with the other Latin American studies. Sorg (1983) showed that an island (Vinalhaven, Maine), formerly thought to be highly inbred, has the relatively low inbreeding coefficient by isonymy of 0.0034 (0.0017 after correction for polyphyletic surnames) and an even lower coefficient from pedigree analysis of 0.0003.

Another island study is that of Ramea Island off the coast of Newfoundland (Devor *et al.*, 1984). A fish processing plant was built on the island in 1950–2 and opened up what had been a small, isolated fishing port to economic expansion and immigration. Most of the married individuals studied there were not born there. Among 201 marital pairs, 25 were isonymous (compared with 5.15 expected at random); so F = 0.0315, F_r = 0.0064 and F_n = 0.0253. The low ratio of the random to the non-random component is explained in part by the fact that 16 of the 25 isonymous matings were between individuals from elsewhere – in general, the little villages up and down the coast whose genetic structure is presumably more inbred that that of Ramea today. Thus the third most common surname in the Ramea marriages yields two instances of isonymy but occurs in only one individual who was born there – the daughter of two immigrants of the name. In the case of another surname, all nine individuals with it (including an isonymous married couple) are from elsewhere. In combination with data on marital migration and blood polymorphism frequencies, the surname data show that the population of Ramea Island is an amalgam from various isolated fishing villages with a few individuals from farther away. The largest of these other places, Burgeo, and another island, Fogo, are being studied and the population structure compared with that of Ramea (Koertvelyessy *et al.*, 1984). On Fogo Island $F = 0.0111$, $F_r = 0.0042$ and $F_n = 0.0069$. There is a gradient in inbreeding coefficients from Burgeo to Fogo to Ramea and on the small islands the non-random component is larger than the random one so the statistical confidence limits of F may be wide.

Kirkland and Jantz (1977), in a study of the rural population of the same Appalachian Mountain county where Pollitzer and Brown (1969) had examined the Melungeon isolate, found higher values of the inbreeding coefficient than in other US studies, but the non-random component (approximately 0.0052) is again larger than the random component (approximately 0.0047), again suggesting that sampling error may be large or that the population being sampled is an amalgam of smaller, more inbred subpopulations.

Biological inferences from studies of Latin American surnames have flourished. As mentioned earlier, Shaw (1960) saw the special advantage for studies of inbreeding of the practice in Spanish-speaking countries of regularly using two surnames from different ancestral lines: those of the father and of the mother's father.

Yasuda and Morton (1967) explored the properties of isonymy in Brazilian populations, but they were aware of the shortcomings in estimating inbreeding from isonymy there. Azevedo *et al.* (1969) noted that in Brazil there is a large amount of polyphyly of certain devotional surnames such as Jesus that parish priests had repeatedly assigned to different families. For this reason Azevedo *et al.* (e.g. 1982, 1983) confine most of their surname studies in Brazil to the associations of types of surname with racial groups and with genetic characteristics: there is a strong association of devotional surnames (those from saints' names, religious festivals, the Cross, etc.) with Blacks; surnames from names of plants and animals are found where Amerindian origins are dominant; and other surnames, such as those of Portuguese origin, are often associated with the European element in the population. Mulattos in Brazil generally show intermediate frequencies of surname types.

In a series of studies on the north coast of Peru (Lasker, 1968, 1969, 1977, 1978*b*) quite high levels of isonymy and of the coefficient of relationship by isonymy were found between communities and even higher levels occurred within them. These high levels are due in part to the high frequency of certain common Spanish surnames (including devotional surnames) that may have been assigned to different families by the Spanish priests who were 'civilizing' the Indians. Despite this polyphyly (that contributes to higher values of the coefficients of relationship by isonymy than can be accounted for by known migration rates: Lasker, 1978*b*) these studies have been of value in the development of methodology and they have shown the low variability levels of isonymy calculated from different sources of surname data (Lasker, 1969).

Halberstein and Crawford (1975) have applied the method of isonymy to a series of populations of Tlascalan Mexican ancestry and have shown a gradation of inbreeding levels among them. The three communities are: (1) San Pablo, a rural Tlascalan-speaking area isolated on a mountain slope at the southern extremity of the state of Tlascala; (2) the city of Tlascala, the capital of the state; and (3) a transplanted Tlascalan population at Cuanalan near Mexico City. Although marriage with someone from the same small area of the town (*barrio* endogamy) is only slightly more frequent within San Pablo than within either of the other places, those marrying into San Pablo were much more likely to come from nearby places than those marrying into Tlascala or Cuanalan. Marital isonymy was highest in San Pablo (2.16%), next most frequent in Cuanalan (1.84%), and least common in the city of Tlascala (1.30%), but the numbers are small (12 in 556, 5 in 272 and 3 in 231, respectively) and not statistically significantly diverse. Unfortunately the data on random isonymy, which might be more informative, have not been published. It would also be interesting to know whether the coefficient of relationship by isonymy between these Tlascalan populations and between them and other ones in Saltillo and New Mexico are significantly higher than is general among diverse Mexican groups. If this is so it would mean that relationships dating back to the time of the Spanish conquest in the early sixteenth century when the Tlascalans dispersed have left traces in the Mexican population structure that are still detectable in surname distributions. A negative finding on adequate-sized samples would imply the opposite, equally interesting, conclusion.

Two hispanic enclaves in New Mexico (Abiquiu and Rosa) have been studied and reported on by Devor (1980). They have inbreeding coefficients as measured by isonymy of about 0.05. Devor warns that this value represents an excess from isonymy because of the small number of available surnames. Since, at Abiquiu, the non-random component of the inbreeding coefficient was reported to be 4 to 6 times as great as the random component, this explanation cannot account for the larger part of these inbreeding coefficients, however. In a subsequent study (Devor, 1983), matrix methods revealed that the high level of isolation has broken down since 1940 to a greater extent than traditional isonymy and marital distance studies suggested.

Lasker *et al.* (1984) found very high coefficients of relationship by isonymy within two towns in Mexico (Tzintzuntzan and Paracho) and moderately high coefficients between them. This latter finding is largely caused by the presence in both places of high frequencies of certain common Spanish surnames that are probably polyphyletic (Estrada, Morales and Hernandez). However, 113 different surnames are shared by the two towns and these include two surnames of Tarascan Indian origin. Although there was no evidence of recent direct migration between the two places, there is regular commerce and some intermarriage between the Tarascan Sierra, where one of them is located, and the Tarascan Lake Patzcuaro Region which contains the other.

In a study in Laredo, Texas, Weiss *et al.* (1983) found only 6571 different surnames (only 3854 if equivalent names with different spellings are merged) in 140560 baptisms. When this is compared with the 32457 different surnames in 165510 listings of marriage records in England and Wales (Lasker, 1983), it seems likely that polyphyly is more prominent in the Spanish-surnamed Mexican population than, for example, in the population of Great Britain. Devor's (1980) comment that the level of isonymy in a Mexican population in the United States is inflated by the dearth of available different surnames can probably be extended to other Mexican populations and also to Latin Americans in general.

By contrast with the New World, few studies of isonymy (other than those of mountain communities reviewed in the previous chapter) have been undertaken in continental European populations. Morton and Hussels (1970) and Morton *et al.* (1973) extended their studies to the Swiss population as a whole by ingeniously combining the surnames of random pairs of individuals from national surveys with data on the size of communities and the distances between them. Combining the theory that the logs of genetic differences are inversely proportional to geographic distances with the Crow and Mange method for inbreeding by isonymy, Morton *et al.* are led to the conclusion that because of polyphyletic surnames isonymy gives an inflated estimate of local kinship coefficients, but that isonymy is in good agreement with other estimates of kinship at greater distances. They conclude that the inbreeding coefficient for the whole of Switzerland is approximately 0.0009.

There is also information from surnames on population structure in Sardinia. Floris and Vona (1980) studied data on the surnames in 1375 marriages (in six widely separated communes) that took place in 1876–1976. The observed isonymy averaged 1% with a range of 0.3–2.2%.

Since most of the subsamples were small, there is little that one can say about them. The coefficients of relationship by isonymy between communes, which were based on larger bodies of data, ranged from 0.0004 to 0.0051 and imply modest degrees of relationship between communes. Compared with the Italian Alpine communities studied by Kaplan *et al.* (1978), the communes studied in Sardinia are more widely spaced and the coefficients of relationship by isonymy are lower.

Zei *et al.* (1983*a*,*b*) made an even more comprehensive study of surnames in 48470 Sardinian marriages. Calculations were carried out separately for each of the 11 dioceses on the island. The authors estimated the rates of immigration into these by counting the excess of rare surnames over what would be expected in a closed population with surname frequencies dependent on random selection for survival. These estimates of immigration rate correlate well ($r = 0.71$) with the degree of industrial development. The log of degree of relationship between pairs of dioceses is negatively correlated with the distances between them. The relationship is not linear, however; as in other studies, there is more kinship at long distances than the theory would explain. As did the Lasker and Mascie-Taylor (1983) study, Zei *et al.* found the drop-off with distance to be steepest for rare surnames, intermediate for surnames of moderate frequency, and least marked for common surnames. This is ascribed to a greater extent of polyphyly of the more common than of the rarer surnames. The heterogeneity among surnames is one explanation of the departure of the distance relationship from that predicted by the Malécot theory. Another reason for deviation from the theory may be diversity in modes of transportation – longer journeys are made by faster vehicles. If distances were measured in hours taken by the usual conveyances rather than in kilometres, the ratio of log kinship to distance would be more nearly linear.

There has been one other large study in Italy (Cavalli-Sforza and Bodmer, 1971). The proportion of isonymy in the Parma Valley was traced from the late-sixteenth century to the mid-twentieth. There was a rise in isonymy in the first 100 years until the inbreeding coefficient by isonymy reached about 0.025; there was then perhaps a gradual decline to 0.0158 in marriages in the period 1930–50. The study also showed that in marriages between second cousins, the more numerous the male links in the pedigree, the more numerous the marriages: there were 774 marriages between patrilateral second cousins (those with only male links and hence isonymous) but only 252 between matrilateral second cousins (those with no male links in their pedigrees). Most other degrees of relationship showed the same tendency. This is just what one would expect if propinquity is a factor in mate selection and if virilocal residence patterns lead to more migration of females than of males. Whatever the cause, this disruption of the ratios assumed by the Crow and Mange method means that the estimate of inbreeding by isonymy in this population overstates the actual extent of inbreeding. This is in addition to any overestimate caused by duplication of surnames at the time they were first introduced.

V. Fúster (at the III Congreso de Antropología Biológica de España, in July 1983) reported the frequency of isonymy in nine parishes of a *municipio* in Galicia, Spain. The levels are similar among parishes and throughout the span of time since 1871 and yield inbreeding coefficients and values of its random component that are comparable to coefficients derived from the migration matrices but are about 70% higher than rates calculated from dispensations. The three most frequent surnames – López, Fernández and García (20% of all occurrences) – account for 43% of the mean inbreeding coefficient from isonymy of 0.0079.

There have been a few studies of inbreeding by surname models in other countries of Europe also. Some Hungarian populations have been studied and German historical data have been examined by Weiss (1974). In addition itinerant gypsy groups of many European countries are known to be highly inbred (Sunderland, 1982) and this is reflected in the distribution of surnames among them. Avcin (1969) assessed the consanguinity by isonymy among Slovenian groups; an inbred group in Paris had only five surnames and in several groups the majority of members had the same surname. The isolation of gypsies and high inbreeding among them is probably more because of their way of life than because of their ethnic origins. At least itinerant groups with no apparent Romany connection, such as the tinkers of Scotland, have very few surnames among them and are known to be a relatively closed, highly inbred population.

8 *Scotland and Ireland*

The surnames of Scotland, Ireland, Wales and England have somewhat different histories and therefore vary in time of origin and in geographic distribution patterns.

Highland Scotland had clan names rather than hereditary surnames. After the defeat of the clans at the battle of Culloden in 1746, however, the transition was rapid; many Scots in the border region took English-sounding surnames, and Scottish names became more frequent in England. Many Scottish surnames are indistinguishable from Irish ones and the difference between Mac and Mc is of little help. These prefixes indicate that the surnames of which they are a part arose from clan names or patronymics (that is, from a father's given name).

In Ireland the clan names apparently go back throughout historic times. Eventually some were Anglicized and the O' was often dropped. Many names of Norman origin also took root in Ireland. More recent migrants to Ireland were often Protestants and they carried different surnames that are usually distinct from those derived from the old clans. The clan names, at first closely associated with particular counties, spread to other areas but are still sometimes more frequent in their area of origin than elsewhere in Ireland. Names of non-Irish origin were usually introduced into large towns and cities, such as Dublin, and of course names of Protestants from Scotland are most frequent in Ulster.

In Wales patronymics continued in use throughout the Middle Ages. The saints' names were eventually converted into hereditary surnames, retaining the final 's' which means 'son of'. Most of the common Welsh surnames are of this form: e.g. Jones, Williams, Evans, Hughes, Phillips, Roberts.

In England surnames were first used by the nobility and were only gradually adopted by lower social classes. They first became common in the south. In the eleventh century the same person might use alternative afternames in different contexts, but use of hereditary surnames soon became common and they were in general use by the fourteenth century and almost universal by the fifteenth. Spellings were not standard, however, but varied according to local pronunciation or the whim of the scribe. Like Scottish and Irish surnames, some English surnames are recent in their present forms. Only when the function of keeping vital records passed from the church to the state in the nineteenth century did spellings of surnames generally assume standard form. Changes continued to take place, however. For instance Clarke was the thirty-ninth most common surname in 1853 and rose to twenty-sixth, twenty-fifth and twenty-second in three lists in 1975, whereas Clark dropped from twenty-seventh to thirty-sixth, thirty-sixth and thirty-fourth in the same period (Ammon, 1976); the shifts are almost certainly caused by a change in preference for the spelling with an 'e', not a greater fertility of the Clarkes than the Clarks.

Knowledge of the population structure in Scotland from surname distributions is limited to studies of islands off the coast and of two small coastal villages. One of the most thoughtful attempts to use surname analyses in human population biology is the study by Morton *et al.* (1976) of the Isle of Barra in the Outer Hebrides. Barra and the immediately surrounding islets were divided for the study into 14 areas. The island of South Uist, the rest of the Outer Hebrides, the Inner Hebrides, the Scottish Highlands and the rest of Scotland were also studied. None of the separate results were published, however; subsample sizes were probably deemed to be too small for that. For isonymy, surnames were divided into 26 classes with the rarer names grouped by origin and similarity. The McNeils and perhaps the McKinnons and McDonalds have been present on Barra since the fifteenth century and in 1910–39 people of these three surnames still constituted 43.8% of the population. The authors estimated the coefficient of kinship within localities relative to Barra as a whole to be 0.0042 and relative to Scotland as a whole to be 0.0084. However, from the observed marital isonymy (some 10% of marriages) relative to contemporary panmixia the mean inbreeding coefficient (i.e. the non-random component of other formulations) was only 0.0007. Thus internal deviations from panmixia are negligible but the fit with an isolation-by-distance model is good. Historical demography, migration rates and genealogy are consistent with isonymy in defining the population structure – which is said to be similar to that of other British isolates. Inbreeding rates of this order of magnitude are not likely to produce any startling gene frequency changes. This may in part explain the dearth of interest by many human geneticists in questions of this kind and their over-attention to extreme instances of isolation – an over-attention which has tended to produce a distorted image of the breeding structure of the human species.

One other study has been undertaken of surnames in the Outer

Hebrides: that of Robinson and Clegg (1981) and Robinson (1983) on Eriskay. There was little deviation in the extent of isonymous marriages from that predicted as occurring at random. Of 144 marriages since 1861, 11 were isonymous, so the total inbreeding coefficient as calculated from isonymy was about 0.019. Any effort to consider trends over this period is handicapped by the small size of the population, which is itself, of course, a cause of the high level of inbreeding. The values calculated by isonymy are all much higher than those derived from data on dispensations granted by the church (which on Eriskay, as on Barra, is Roman Catholic).

Roberts and Roberts (1983) conducted an interesting test of the isonymy method in samples of the Orkney population. They drew two samples of 75 individuals each and compared the coefficient of relationship by isonymy between them with the coefficient of kinship from the pedigrees. After allowance for different constants, the former coefficient was sevenfold larger than the latter. The difference is shown to be caused by 29 of the 33 pairings with identical surnames not having a traceable relationship in the pedigrees. Among the 66 pairs of individuals traced as related in the pedigrees, the expected isonymy (4.39 cases) is close to the observed number (4). The Roberts find it implausible that pairs of individuals who show no interrelationships by pedigree analyses that are complete for all possible close links would, in fact, have coefficients of kinship averaging as high as 0.001285, the level shown by isonymy. They therefore conclude that the surnames are highly polyphyletic and that where patronymic systems of afternames have given way to hereditary surnames only recently (as in the Orkneys and, even more so, Shetland) studies of relationship by isonymy will be flawed. Some studies show little difference in the degree of kinship revealed by isonymy and from pedigrees, but the ratio of 7 : 1 in the Orkneys and about half that amount in Saas, Switzerland (Hussels, 1969) and Kippel, Switzerland (Ellis and Friedl, 1976) lead the Roberts to prefer to limit use of isonymy to within-population estimates of relationship and to relative rather than absolute levels.

Bailie (1981, 1984) studied the surnames of the residents of two villages 25 kilometres apart on the Moray Firth. From the occupations of bridegrooms given in the records of marriages in the period 1855–1975, she divided the population into fishermen and 'others' ('others' include a variety of service occupations). Among the fishermen there were very few surnames and these tended to be of high frequency, so there were more isonymous marriages among fishermen than among non-fishermen. However, the fishermen had few surnames in common between the two villages although there was apparently somewhat more intermarriage between the two villages involving fishermen's families. This suggested that there may have been a certain degree of isolation of the fishing communities of the two villages, but their coexistence with those of other occupations is one of the factors that have made the society more 'open' in terms of its genetic structure than otherwise would have been the case. The use of isonymy to study social stratification through inheritance of occupational specialization is novel. Besides these studies of Bailie in Scotland, Smith and Hudson (1984) and Smith, Smith and Williams (1984) are studying the surnames of occupational subgroups in the coastal parish of Fylingdales in North Yorkshire and report a similar finding of differentiation between the families of fishermen and those of men in other occupations.

In Ireland, Relethford (1982) used the concept of conditional kinship (essentially twice the coefficient of relationship by isonymy) to study random isonymy among seven largely endogamous isolates within a compass of 107 kilometres of each other in Galway and Mayo counties. The rank order correlation of kinship and distance was −0.782. Since Relethford found a corresponding correlation of only −0.29 in the general population of rural western Ireland, he argues that distance is a more important influence on kinship among isolated communities of fishermen and small-plot farmers than in more heterogeneous populations in which isolation is not so great.

Relethford, Lees and Crawford (1981) analysed data from 12 towns in rural western Ireland (seven in County Clare, two in Galway and three in Mayo) and estimated conditional kinship. The detailed results are not published but were analysed as a function of geographic distance between towns and gave a significant fit with the Malécot model of isolation by distance. Western Ireland is described, on this and other evidence, as typical of continental regions that show a low degree of local isolation and a moderate decrease in kinship with geographic distance. Deviations seem to be due to the sizes and economic statuses of the towns.

A further study of these populations has been published by Relethford and Lees (1983). Surprisingly, between-town isonymy based on surnames of males is not correlated with that based on females. The samples were probably too small for reliable results, however, the average sample size being 29 for males and 22 for females. Thus the median was probably less than 20 and one would expect only one or two instances of isonymy in most between-town relationships; chance factors would play a very large role. Despite the small numbers, however, female isonymy is correlated with female anthropometric differences between towns and with the physical distances between towns. These correlations would appear to be statistically significant, but as the authors themselves imply, the number of degrees of freedom is exaggerated in a correlation of sets of differences because of intercorrelations within each of the two matrices being compared.

9 *Regions of England*

North of England

The population structure of part of Northumberland has been studied by the method of surname analysis. Dobson (1973) examined the parish records of four ecclesiastical parishes that lie along the valley of the River Coquet, and found that the retention of surnames in the population was similar in the four parishes. Of the surnames present in 1780–1809, 47–53 % had been present in the same parish in 1720–49 and 39–41 % had been present in 1690–1719. The further apart a pair of parishes, the fewer surnames they shared. Analysis of the surnames plus an examination of marital distances led Dobson to conclude that the population structure, and amount and distance of gene flow, led to effective population sizes too large to permit local genetic differentiation. Nevertheless, there was much less gene flow into these parishes than into that of Horringer in East Anglia where of the surnames present at the end of the eighteenth century only 2 % had been present in the latter part of the seventeenth century (Buckatzsch, 1951).

One of the Northumbrian parishes, Warkworth, was further studied by Rawling (1973). After merging different spellings of the same surnames there were 22 instances of isonymous unions in the period from 1686 to 1812. Eighteen of these were endogamous marriages and four exogamous marriages. The amount of inbreeding, and especially the size of the non-random component, increased with time. However, the random component actually decreased, although perhaps not statistically significantly. The author concludes that the inbreeding coefficient (0.0056 to 0.0072, depending on assumptions), which is as high as is found in isolates, is not likely to be correct for this open population with its economy based on farming and the raising of sheep. Roberts and Rawling (1974) extended these studies to the three other parishes and conclude that the trends in inbreeding shown by marital isonymy appear to correlate with historical events. However, the estimate of total inbreeding, and hence that of the non-random component also, is subject to considerable sampling errors – especially in the subsamples with only two to nine instances of isonymy in each – and the estimate of random inbreeding (that is, the accumulated effect of past matings) is more reliable. These values (ranging from 0.00078 to 0.0014) are less variable and lower and hence a better estimate of the breeding structure of this parish. Roberts (1982) concludes from the same surname data that there was restricted migration, at least of the males (female movements are not reflected in surname distributions), and that the pre-existing bondage system was one reason residence remained permanently at or near birthplace. Roberts showed that the sizes of these local populations were large enough and that there was sufficient migration to prevent random differentiation from each other by genetic drift.

Roberts (1980, 1982) also studied a parish across the River Tyne from the city of Newcastle. Whickham was initially a rural community that developed industrially with river traffic, coal mining and an iron works, and is now a suburb of Newcastle. Of 5780 marriages between 1579 and 1886, Roberts found 48 were isonymous. However, when the time span was divided into 13 periods of 25 years each, there were only one to seven isonymous marriages per period, and too much emphasis should not be placed on differences between periods in the total inbreeding coefficient or its non-random component. The random component is more reliable and ranged from 0.0022 to 0.0056. These values, calculated from marriage records, tended to rise at first and then declined in the last three periods surveyed (and, no doubt, over the last century also). The inbreeding coefficients are much lower than those of the parishes in the Coquet valley.

Lasker and Roberts (1982) studied the baptismal records of the same parish. After correction for the different constants of the respective formulae, the values for the coefficient of relationship by isonymy within a time period should correspond with the random component of the inbreeding coefficient calculated from marriage records some 20-odd years later. This was not apparent in our results. Instead we found an early drop in the coefficient of relationship and only small changes later. Thus even the random aspect of surname commonality may vary with the source of the data.

Another English north country parish that is being studied is Fylingdales in North Yorkshire (Smith and Hudson, 1984; Smith *et al.*, 1984). The fishermen, mariners and ship-owners and their families form a more or less separate subpopulation from farmers, agricultural labourers, alum works labourers and others. The degree of separation decreased with time, as apparently did the extent to which sons followed their fathers' occupations. The extent of intermarriage between individuals whose families were engaged in different trades should be tested before surname differentiation by trade can be taken as evidence of community subdivision, however. The Fylingdales study ingeniously reduces the complexity of the numerous comparisons of pairs of occupations by an interesting method called multidimensional scaling. This two-dimensional image of the interrelationships shows the differences between occupations to be decreasing and the diversity within them tending to increase.

Otmoor

Another region of England where the structure of the human population has been studied by surname distributions as well as by other methods is that of the villages surrounding the Otmoor (Lasker, 1978*a*; Küchemann *et al.*, 1979). The Otmoor is about 1600 hectares of uninhabitable, low, flat, sticky mud surrounded by villages whose inhabitants earned a living, as smallholders or farm labourers, supplementing their income by grazing pigs, hunting fowl and digging peat on the moor. Those from the different villages stood together in trying to resist enclosure of the moor and until after the mid-nineteenth century they tended to intermarry among themselves. Marriage records from 1753 to 1950 of eight parishes around the moor yielded information on 6142 individuals. Coefficients of relationship of isonymy (*Ri*) between parishes and between them and outside areas were calculated in various ways. Whether or not different spellings of what were thought to be the same surname were merged made some difference to the level of the coefficients, but results by the two methods were highly correlated ($r > 0.9$). Data on the separate sex combinations were also significantly correlated, but not so highly ($r = 0.4$–0.8). The mean *Ri* (spellings merged) averaged 0.0093 within separate Otmoor parishes, 0.0016 between them, and 0.0007 between them and more distant places.

A previous study of the surnames of males in the lists of those eligible to vote in 1976 in these and surrounding parishes found that *Ri* (spellings not merged) averaged about 0.00075 between electoral parishes and about 0.0005 between them and the nearby city of Kidlington (Lasker, 1978*a*). From the two studies it is seen that the coefficient of relationship by isonymy (and hence also the random component of the inbreeding coefficient) both within and between the Otmoor parishes increased over the two centuries studied but has now decreased. The coefficient of relationship between pairs of parishes and the amount of material migration between them are correlated: $r = 0.53$ in 1753–1850, 0.31 in

1851–1950 and 0.75 in 1976. The most recent period shows the highest correlation because current coefficients of relationship incorporate the effect of the very migration that has been enumerated. *Ri* correlates negatively with the geographical distance between the parishes: $r = 0.36$, -0.23 and -0.27 for the 1753–1850, 1851–1950 and 1976 periods, respectively.

Fenland villages in Cambridgeshire

A group of five Cambridgeshire Fenland villages (Manea, Doddington, Wimblington, Pymore and Coveney) is the subject of another study (Lasker and Mascie-Taylor, 1983). The land here is low, flat and wet as in the Otmoor, but the Fenland has been drained and is now some of the most productive agricultural land in England. The source of the lists of surnames is the present-day registers of electors and the purpose of the study was to determine the population structure that had accumulated since the villages were established or since the surnames had become hereditary (both presumably mediaeval events except for subsequent migration into the area and occasional changes of surnames). To study the wider structure, these lists of surnames were compared with those of other places within 40 kilometres, with those of a zone 40–65 kilometres away, and with those of all England plus Wales. Among the villages the coefficient of relationship by isonymy averaged 0.00076, almost exactly the same as among Otmoor villages. With the zone within 40 kilometres of these villages the weighted average *Ri* is 0.00054, with the zone 40–65 kilometres away it is 0.00045 and with all England and Wales it is 0.00042. These values are also in keeping with the results from the Otmoor region, where more distant places gave an *Ri* of 0.0005. The Fenland villages show a negative correlation between distance and *Ri* ($r = -0.55$), a feature previously seen in the Otmoor communities and clearly also true for the Coquet valley in Northumberland, to judge by the numbers of surnames in common in different pairs of parishes.

In the Fenland villages surnames were separated into different classes according to the frequency of their occurrence in all England and Wales. The rarer a surname, the more likely that if it does occur in one of these villages that other occurrences will tend to be nearer. The problem is that while common surnames make by far the largest contributions to surname commonality, the rare surnames are more informative about relationship. The advantage of dividing surnames into frequency classes – if sample sizes are large enough to warrant it – is that rare surnames are less likely to be polyphyletic and therefore provide more direct evidence about local migration. Common surnames, on the other hand, tend to probe a wider area and a deeper past but surname commonality has a less clear genetic basis and use in genetic models must therefore be on a relative rather than an absolute scale.

Other regional studies

Souden and Lasker (1978) examined the surname distributions among 37 parishes in East Kent from data in the Marriage Duty Act returns of 1705. Unfortunately most of the parishes had few residents (ten of them had fewer than 20 adult males each) so the sample sizes are too small to permit interpretation of specific values of the coefficient of relationship by isonymy in terms of known geographic details or historical events. Among the 351 pairs of parishes with 20 or more adult males each, the mean value of *Ri* (different spellings of the same surname merged unless there was doubt about identity) was 0.00092 and the correlation with distance between parishes was $r = -0.29$. The data from the smaller parishes tended to confirm these values: an additional 114 values yielded *Ri* of 0.00129 and a correlation of *Ri* with distance of $r = -0.26$.

A repeat study of the largest of these East Kent parishes, Ash-next-Sandwich, showed that common surnames in the register of electors of 1976 were no more similar in frequency to the findings of 1705 than if they had been picked at random from the population of all England and Wales. Such complete panmixia was not shown by rare surnames, however. The rare names showed considerable local continuity: four of those which occur in the parish both in 1705 and in 1976 are so rare elsewhere as not to appear at all in a list of over 165000 individuals from all England and Wales.

Using another method, Watson (1975) studied the surname distributions in a number of parishes in southern Cambridgeshire. He used records from 1538 to 1840 and reported the numbers of surnames in common between pairs of communities. Results from the two methods cannot be compared, however, as unlike the coefficient of relationship, the numbers and proportions of names held in common are a function of sample size. The same problem applies to studies of turnover in family names. Coleman (1984) has summarized the various English studies in which the temporal constancy of surnames is reported. Small differences in the proportion of surnames of one period surviving until another may well be due to differences in the sizes or changes in sizes of the parishes studied. However, substantial differences are not thus explained and must represent real differences in the in-migration patterns. Thus Buckatzsch (1951), apparently the first to use such an index, found that 35 % of the surnames in Shap, Westmorland, at the end of the eighteenth century had been present about 100 years before, but that for Horringer, Suffolk, the comparable figure was 3% and, in a slightly longer period, none.

In a study of the distribution of surnames within villages conducted on five electoral parishes just south of the town of Newmarket, Suffolk, and in the five villages in the Cambridgeshire Fens previously mentioned, the coefficients of relationship between parishes in the first set ranged from 0.0006 to 0.0028 (unpublished) and in the second set from 0.0002 to 0.0010 (Lasker and Mascie-Taylor, 1983); these are typical values for the English countryside. However, the average coefficient of relationship between houses with adjacent numbers on the same street was 0.00742, between other houses on the same street it was 0.00249 and between other houses in the same parish it was 0.00158. Thus by this measure the residents of adjoining houses are three times as closely related, on average, as the residents of other houses in the same street, who are about 50 % more closely related to each other than to other persons in the parish who are, in turn, about three times as closely related to each other on average than to the whole population of England and Wales (Lasker *et al.*, 1984). In farming country and villages the localization of surnames has almost certainly been maintained, in some cases, by the partition of property among male offspring. In other instances a 'granny cottage' as residence for an elderly person is established in the immediate neighbourhood of the home of an adult son, daughter, niece or nephew.

Virtually all studies of human population structure list marital migration within the same community as zero or as some arbitrary short distance such as 1 kilometre. Coleman (1979), however, attempted exact measurements even of short marital distances, and the preliminary impressions from studies of the villages south of Newmarket and other areas is that such a practice might yield useful information bearing on the inheritance of land as well as having implications for biological interrelatedness and inbreeding. In the villages south of Newmarket there is a negative correlation between their distance apart and the coefficient of relationship ($r = -0.28$); although the finding is not statistically significant it is in keeping with findings elsewhere. The shape of the curve of the plot of intercommunity *Ri* versus distance may not be linear (although the substitution of log distance for distance does not always give a better linear fit), but the addition of inter-individual commonality of surname and interdwelling distances when these are short, would improve our knowledge of one part of this curve.

Island populations

In England and Wales there are few if any populations as isolated as some of those which have been studied in Scotland and Ireland and there are none as inbred as the true isolates discussed earlier. The Isles of Scilly, 38 kilometres off Land's End, are about as inaccessible as any English community. They yield the highest level of isonymy of any English area (Raspe and Lasker, 1980), but they are on the sea route between France and Ireland and at the mouth of the transatlantic route from the English Channel, so have long had occasional visitors and immigrants by sea. By the eighteenth century, boats from the English mainland were normally arriving at 4 to 6 week intervals. There has also always been plenty of travel by small boat between the islands in the group. In 1871, persons in the islands who had been born there were married to another person from the same island in 59 % of cases and to a person from a different island in the group in another 27 % of cases, leaving 14 % married to outsiders. The inter-island coefficient of relationship by isonymy in 1851 ranged from 0.0022 to 0.0428 with a mean of 0.0119 – a high value compared with other populations and 26 times as high as the relationship of the island population with the population of England and Wales in general. There is a high correlation of inter-island rates of relationship in 1851 and 1871 (r = 0.93) as one would expect, and the correlation of degree of relationship with distance between islands (measured between the nearest docks) is negative ($r = -0.5$). The chief influence on the surnames of those listed in the 1851 and 1871 censuses was evidently the founder effect. The surnames are from Wales and England (predominantly from Cornwall and Devon) and most of the chief ones were introduced in the years following 1571 when the leaseholder of the islands, Sir Francis Godolphin, brought to the Scillies soldiers for his garrison and tenants for his farms.

Raspe studied marital migration as well as isonymy and found that the marital migration evident in the census data was not consistently correlated with the coefficient of relationship by isonymy between the same pairs of island populations. This is not very surprising: marital migration over a single generation should correlate negatively with the non-random component of the inbreeding coefficient (inbreeding current at the time of the study). The coefficient of relationship by isonymy, on the other hand, is identical to the random component of the inbreeding coefficient (except for a different constant) and therefore incorporates the effects of migratory events since the founding of the population or the adoption of surnames: a correlation with migration would be expected only if the migration were studied over that same span of time or the pattern of inter-island marital migration had remained very constant. In the Isles of Scilly study, the marital migration rates apply to movements during a limited period of the nineteenth century, whereas the coefficient of relationship encompasses movements during the same period but also during the whole of the seventeenth and eighteenth centuries and, especially, the period of rapid expansion of the population in the late sixteenth century. If the specific rates of inter-island migration (especially the migration of males) were known for that whole span of time, they would be expected to correlate highly with coefficients of relationship by isonymy calculated on the basis of data from the mid-nineteenth century.

Gottlieb is at present engaged in a study of the genetic structure of the human population of the Channel Island of Sark. Her studies will include an analysis of surname distributions on this and and other Channel Islands. There will then be information on all the more isolated corners of the British Isles – Irish isolates, the Hebrides, the Orkneys, the Scilly Isles and the Channel Islands – and the effect on population structure of their rather modest degree of isolation should be well understood. It is already clear, from both surname data and other types of information, that British populations have been quite open for a long while. The coefficient of relationship by isonymy is positively correlated with migration if the latter is summed over many generations, and *Ri* is negatively correlated with distance between and also within villages, which reflects a pattern of past migrations in which short distances were common, but longer ones also occurred.

10 *English cities and the general population of England and Wales*

Coleman (1979, 1980*a*, 1982) has done a great deal to shed light on the population structure and marriage pattern of a modern English city and its environs. Most of the population of England is in urban areas, but almost all the other studies of population structure have been of villages. Coleman analysed the pattern in Reading, Berkshire, and its environs by a study of all the marriage records for a 12-month period in 1972–3, using copies of the registration records in the districts of Reading, Wokingham and Henley-on-Thames. A questionnaire seeking much additional information was circulated to a sample and returns were received from 946 of the 2396 couples married during the specified period. The average couple had been born 103 kilometres from each other – about the same distance as their parents. The average distance from their place of residence when single to the place where they met their future spouses was small, however, compared with the mean distance from the meeting place to the place where the couple resided after marriage. Thus migration after, or at the time of, marriage was an important element in inter-generational migration. As Coleman puts it, the people become individually more heterogeneous but collectively more homogeneous with regard to ancestry (and hence genetic constitution). There is enough migration that in only three generations the population of the built-up area of Reading would be 95 % homogeneous. For the area as a whole, 95 % of any genetic heterogeneity would be lost in six generations at most because of the flushing through the system of large streams of migrants, in and out. Lower social classes are somewhat more stable geographically than upper ones, but there is enough social mobility for the overall influence to be small. These rates of increase in homogeneity in Reading are somewhat more rapid than Hiorns, Harrison and Küchemann (1973) observed in two parishes of the city of Oxford (St. Giles and Cowley St. John), but much more rapid than recorded for the Otmoor villages. The slowest rates so far reported in England are among the Isles of Scilly (Raspe, 1983), where in some periods it would have taken over 30 generations for two of the islands to reach 95 % identity by the marital exchange rates in effect.

Coleman (1980*b*) has also studied inbreeding in the Reading population. He found that of 2396 marriages, 2 were truly isonymous, giving an inbreeding coefficient of 0.000209. In the subsample of 946 marriages about which responses to the questionnaire were received, 8 were consanguineous, giving an average inbreeding coefficient of 0.000174. The two cases of isonymy were not among these but there was one instance, not counted as isonymous, in which the spouses' surnames are pronounced in the same way but spelt differently and another instance of related spouses in which the surname of one spouse differed from that of the other by the addition of a prefix. In 16 instances a wife had taken her husband's surname prior to marriage; these were not counted as isonymous because in all instances where a woman's name differed from that of her father, the latter's surname was substituted for her own in the calculation.

Lasker *et al.* (1979) further studied the surnames in these marriages at Reading and its environs. After dividing the Reading area into eight geographic districts and adding a ninth category for individuals from outside the area, coefficients of relationship by isonymy were calculated. These averaged 0.00049 between districts and 0.00057 within districts. The former figure is almost exactly the same as for the Otmoor villages with a wider group of communities, for the Otmoor villages with the city of Kidlington, for the wards of Kidlington with each other, and for five Cambridgeshire Fenland communities with other places within 65 kilometres of their centre. Thus in England today $Ri = 0.0005$ seems to be the general level found in several places.

If one takes the 2920 males from the 17 electoral parishes in Oxfordshire and Buckinghamshire that are nearest to the Otmoor and calculates a coefficient of relationship by isonymy with the 4794 persons listed in the marriage records of the three registration districts in northern Berkshire and southern Oxfordshire, $Ri = 0.00049$, the same figure as between regions in the latter area, so this level may hold even at somewhat greater distances. The 50 most common surnames (Ammon, 1976) contribute 0.00033 to Ri between the Otmoor and Reading areas, so it is the common surnames that vary least from place to place and contribute much to a general homogeneity in levels of relationship – a homogeneity which would have been expected from the high levels of internal migration in industrialized England.

A further study of the surnames in Reading and surrounding areas was undertaken by Fox and Lasker (1983). The separate regions and the total sample have distributions of frequencies that form discrete Pareto (zeta) curves. A plot of the log of occurrences of a surname versus the log of the number of surnames with that number of occurrences forms a straight line. However, the slope of the line varies with the size of the sample. Nevertheless, to the extent that surname frequencies fit this regular pattern, it is possible to estimate the variance of the distributions. The standard deviation of a coefficient of relationship by isonymy would be of the order of 25 % of the mean. By this standard, the values of *Ri* within the different districts around Reading could well vary by chance alone: the differences among the districts are not statistically significantly diverse.

The findings from the surnames of the persons married at Reading, Wokingham and Henley are supported by a study of the surnames of everyone married in England and Wales in January, February and March 1975 (Lasker, 1983). The list of these individuals was first edited to eliminate instances where a person was listed more than once by the same surname. Brides listed by the same surname as their bridegroom were also eliminated from consideration, because as Coleman (1980*b*) showed for Reading, cases where a bride has already changed her name to that of her future husband are far more frequent than true cases of isonymous marriage. The list of persons married in this 3-month period is thought to be a good random sample of the surnames of the breeding population of England and Wales; for instance, the number of occurrences of the common surnames on the list correlates highly with the numbers of persons of the same surnames who were born and who died during the same 3 months ($r = 0.995$ in both cases). For all England and Wales the coefficient of relationship by isonymy is 0.00048. This includes both within-community and between-community pairs, of course, but the latter must predominate by far. Of the 32457 different surnames, the 182 most frequent contribute 0.00043 to the coefficient of relationship; that is 0.5 % of the surnames account for 90 % of the level of the coefficient. The surnames Smith and Jones alone account for 30 % of the coefficient. A second case of one of the rarest surnames is about nine times as likely as a second case of one of the common ones to be in the same district as the first, but these most localized surnames contribute almost nothing to the overall levels of the coefficient of relationship by isonymy. On the contrary, levels of the coefficient will be largely determined by surnames that are widely distributed and differences between coefficients are more likely to depend on random variations in the frequency of common surnames such as Smith than on such meaningful events such as the concentration of (generally rare) surnames of immigrants in selected cities. Thus if one were making a study of population structure in Wales from the distribution of surnames (as far as I know, none has been attempted) one would expect higher coefficients than in England because a higher proportion of the Welsh population has common surnames (Jones and Williams, for instance) than is usual in England. The possibility of such findings being based on the history of surnames rather than on the genetic structure of the populations must be borne in mind.

The different implications of surnames of different frequencies is well exemplified by a class of surnames that represents one of the extremes. Each combination of a hyphenated surname is extremely rare and almost always monophyletic. B. A. Kaplan (unpublished) has made a study of their occurrences in England and Wales and finds a variety of reasons for their origin and retention. Most of those now in use date from the eighteenth century onwards. The process of formation continues – among the marriages of the Reading study, three individuals had added a new hyphenated part to a father's surname and one had dropped half of a father's hyphenated surname. In the marriage records from all England and Wales (after elimination of some hyphenated names such as Spanish ones which are probably not hereditary) there were a few (very few) instances of two occurrences of the same combination. After further eliminating the instances where the two individuals with the same hyphenated surname were bride and bridegroom (since most such cases would have been formed by the couple themselves) the majority of the remaining instances listed the two individuals in the same registration district. This is an extraordinary level of localization of occurrence – even by the standard of very rare surnames. Most such instances must represent a pair of closely related individuals and most hyphenated surnames must be monophyletic.

11 *Specific surnames in Great Britain*

Most of the discussion so far has dealt with the whole array of surnames in populations. However, as noted above, there are some differences in implications of those surnames which are rare versus those which are common. Both groups are made up of individual surnames, however, and some aspects of the surname model of population structure may be better understood by viewing surname distributions separately rather than *en masse*.

For instance, suppose there were a surname, say Adams, and Adamses occurred in three places, A, B and C, in 4, 2 and 1 % of the populations respectively (Table 2). The contribution of Adams to random isonymy would be the square or the product of these percentages: 0.16 % in A, 0.04 % in B, 0.01 % in C, 0.08 % between A and B, 0.04 % between A and C and 0.02 % between B and C. Intrapopulation isonymy would average more than interpopulation isonymy: 0.07 % versus 0.047 %. Also, because B and C have fewer persons named Adams than A, they must

Table 2. *Contribution to isonymy of three surnames (Adams, Brown and Cape) in three hypothetical communities (A, B and C)*

	Adams	Brown	Cape
People in each community with each surname (%)			
A	4	1	2
B	2	4	1
C	1	2	4
Contribution to isonymy (%)			
A × A	0.16	0.01	0.04
B × B	0.04	0.16	0.01
C × C	0.01	0.04	0.16
Total within community	0.21	0.21	0.21
A × B	0.08	0.04	0.02
A × C	0.04	0.02	0.08
B × C	0.02	0.08	0.04
Total between communities	0.14	0.14	0.14

have higher frequencies of some other surnames. If three surnames, Adams, Brown and Cape, each occur in 4, 2 and 1 % of the population in a different one of the places A, B and C, the total contribution of these three surnames would be 0.21 % to intrapopulation isonymy in each of A, B and C, and 0.14 % to interpopulation isonymy in A by B, A by C and B by C. The greater degree of relationship within one's own group than with members of some other group (whether in respect to surnames or genes) follows from the fact that one is more likely to have a surname or an allele that is common in one's group and everyone else in the group is also more likely to have the same genes or surname than are members of a different population with somewhat different frequencies of genes or surnames.

In the hypothetical example, and with reference to the three surnames, the three communities A, B and C have identical degrees of random inbreeding. However, one intuitively knows that the distribution of persons called Cape has a different significance from the distribution of persons called Brown. Cape is a rare surname whereas Brown is ubiquitous. Therefore one will be more struck by the fact that there are many persons called Cape at C and that Cape occurs at all three places, than by the fact that the surname Brown is frequent at place B and also occurs at A and C. The geographic distribution of some particular surnames will therefore now be considered.

One class of surnames comprises the largest number of different surnames, but almost all surnames of the class occur infrequently: surnames derived from place names. These names lend themselves to the study of surname localization. Many of the cities that gave their names to English civil registration districts have also given them to surnames. A search of the list of marriages in England and Wales in the first 3 months of 1975 shows that such surnames were more common in the district which putatively gave rise to the surname and in the five other nearest registration districts than one would expect by chance (Lasker and Kaplan, 1983). There were 89 occurrences of the corresponding surname in the district of the name or in one of the five other nearest districts; this compares with 51 that would be expected at random. A larger follow-up study of some of the same surnames in telephone directories shows that among 13 109 cases of the selected surnames, 328 (compared with 172 expected at random) occur in the telephone exchange area of the place with the name (Kaplan and Lasker, 1983). Thus there are about 90 % more occurrences there than expected by chance. Of course, after the some 25–30 generations since most of these surnames were established, most persons with them live somewhere other than the places where they originated. Nevertheless, the extent to which English surnames have been retained near their place of origin is underestimated by these studies. First, many of these surnames have multiple origins – there has been more than one place of the same or similar name or someone who had no direct connection with the place took the name. Secondly, place-name surnames generally arose when individuals moved away from the place and these moves would more often have been to distant places in the case of town and city dwellers than for villagers. Initially most migrants with place-name surnames came from villages within about 30 kilometres of the place in question (McClure, 1979), but the surnames soon spread – though remaining more common near their place of origin (McKinley, 1976).

In order to determine more about the general distribution of surnames, another group of surnames has been studied: those of frequent occurrence in the list of 1975 marriages. Because they occur more frequently than other surnames they are more comparable in frequency to polymorphic genes (genes with variants of common occurrence such as blood groups). The distributions of surnames may therefore be compared with the gene frequency distributions of blood groups.

Most studies of the geography of genetic polymorphisms have been merely descriptive and yield maps showing somewhat simplified pictures of the distributions. In some of them clines (gradual transitions in frequencies) are shown. However, some efforts have been made to apply theories of diffusion to such maps. Morton (1973), for instance, has applied the Malécot theory of isolation by distance to gene frequencies.

Using this model, Imaizumi (1974) tested the distribution of ABO and Rh(D) blood groups in Great Britain. He also calculated the regressions of the frequencies of these genetic polymorphisms against latitude and longitude and in that way demonstrated the existence of clines. Using specific surnames this work was carried a little further. The regressions against latitude and longitude are at right angles to each other, hence the amount of variance explained by them is additive and the slopes of the two regressions determine the plane which best fits the data. The residuals (deviation from the plane) are also of interest and the amount of variation explained can be increased by polynomial regressions, that is, the relation of the square and cube of the frequency, as well as of the frequency itself, to latitude and longitude.

A first study of this kind was of the names Smith and Jones (Mascie-Taylor and Lasker, 1984). These two surnames provide the largest samples and their history and distributions are reasonably well known. Smith was already recorded in Anglo-Saxon times and Jones by the thirteenth century; Smith derives from the occupation of blacksmith and

Jones means the son of John (Reaney, 1976). Guppy (1890) reported that among land-owning farmers the Smiths occurred in all parts of England but were least frequent in the south-west and in Wales and were most common in Gloucestershire, Warwickshire and Staffordshire in the west and Essex in the east. Thus they were most numerous north of the latitude of the Thames and south of that of the Wash. Guppy described Jones as the most characteristic Welsh surname, being especially frequent in North Wales; in England he found it common near the Welsh border and in Northamptonshire, Buckinghamshire, etc., but not in the northernmost counties.

From counts of the number of individuals of each of these two surnames who were married in each registration district of England and Wales during January, February and March 1975, the approximate locations of the 1773 Joneses and 2198 Smiths have been plotted on maps (see Appendix). The expected number in each district (under the assumption of equal frequency in all places) was also calculated. The difference between observed and expected frequency was divided by the expected frequency to give a measure of relative frequency. Then the values of relative frequency were regressed against latitude and longitude to describe the form of the geographical clines and to estimate the amount of variance explained by the clines. None of the fourth and fifth power terms of the polynomial regressions was statistically significant, so only the linear, square and cubic terms were retained in the analysis. The frequency of the Smiths increases steadily towards the east whereas the Joneses reach their maximum at the longitude of Wales and their minimum at the longitude of London. In the north–south direction both surnames reach their maximum relative frequencies about 100 kilometres north of London and decrease to the north and south. These polynomial regressions are highly significant statistically and account for 11.8 % of the variation in relative frequencies of Smiths and 36.4 % of that of Joneses, and in both cases the strongest single component is the linear east–west regression. These two surnames undoubtedly had polyphyletic origins, but of course by the time they were established, all the blood group polymorphisms also had multiple foci of high frequency, so the model is appropriate for that kind of genetic entity. On the basis of the model one would expect neutral-polymorphisms (those in which the different forms of the gene have the same effect upon the likelihood of leaving offspring) to exhibit broad clines over hundreds of kilometres. Although the clines would account for appreciable fractions of the variation, still more of the variation would remain unexplained.

These studies have now been extended to 49 other surnames (G. W. Lasker and C. G. N. Mascie-Taylor, in preparation, and see Table 3). Most of these are English and Welsh surnames that occur in high frequency, but a few Scottish, Irish and Indian surnames were included to see whether they had special patterns of distribution. In this study, the difference between observed and expected numbers at each of the 443 registration districts was divided by the square root of the expected number. This kind of relative likelihood of frequency deviation is used in preference to the simple ratio because more of the variance can be explained thereby. Correlation of these values between each pair of surnames and regression of them against latitude and longitude show that of the 49 surnames, all but 17 have a statistically significant relationship to latitude or longitude. Seven surnames which can be thought of as Welsh show a strong tendency to decrease from the west towards the east and 11 were prominent towards the north or the south. Between pairs of surnames all the high correlations are positive and factor analysis shows that Welsh surnames (Davies, Edwards, Evans, Griffiths, Hughes, Morgan and Morris) group together in one group; Indian surnames (Patel, Mistry and Kaur) form another; and there are somewhat less distinct groupings of Irish names, of some northern names, and of some southern ones. In the south and the north there is considerable independence indicating that some surnames are frequent in certain northern districts whereas others are frequent in different northern districts, for instance. There are a number of factors that isolate a single surname, which implies more or less unique intercorrelations for several of the names.

A still more recent study (C. G. N. Mascie-Taylor and G. W. Lasker, in preparation) analyses the distribution of the 84 most frequently occurring surnames. For this set of surnames the degree of commonality ranges from 0.17 to 0.85 of the geographic variance. Thirty factors explain this commonality, the two most prominent of which account for over one third of the variance. All 20 Welsh surnames on the list are high in one or the other or both of these first two factors and no non-Welsh surname is high in either of them. Names common in South Wales form the first factor; those common in North Wales tend to be high in the second factor. The third factor groups together seven surnames that end in 'son' and one other surname, Bell. This third and also some other factors group various 'son' surnames with each other and with other surnames of generally north of England prominence.

Regressions of the values for the common surnames against latitude and longitude (see Appendix) reflect the generally higher frequencies towards the west of the 20 surnames high in the first two factors and identified as Welsh. Among the English, as contrasted with Welsh,

Table 3. *Correlation of likelihood of frequency deviations of surname with longitude (easting) and latitude (northing)*

Surname	Cases[a]	Pearson r^b	
		Easting	Northing
Adams	243	0.04	−0.17**
Allen	332	0.10*	0.05
Anderson	226	0.05	0.09
Bailey	285	0.05	−0.01
Baker	398	0.08	−0.25**
Barker	218	0.06	0.22**
Bell	257	0.02	0.35**
Bennett	334	0.02	−0.06
Campbell	174	−0.03	0.10*
Carter	228	0.04	−0.07
Chapman	237	0.03**	−0.04
Clark	351	0.19**	0
Clarke	417	0.06	0
Collins	261	0.04	−0.23**
Cook	271	0.10*	−0.04
Cooper	387	0.05	0.03
Cox	272	0.10*	−0.15**
Davies	867	−0.39**	−0.05
Davis	310	−0.02	−0.23**
Edwards	445	−0.23**	−0.06
Ellis	222	−0.08	0.01
Evans	701	−0.39**	−0.05
Foster	273	0.02	0.19**
Green	473	0.13**	0.06
Griffiths	333	−0.39**	0.01
Hall	467	0	0.24**
Harris	438	0.03	−0.24**
Harrison	369	−0.02	0.28**
Hill	403	−0.03	−0.04
Holmes	205	0.01	0.10*
Hughes	512	−0.31**	0.08
Hunt	225	0.13**	−0.13**
Kaur	270	0.07	−0.02
Lee	306	0.05	−0.02
Martin	396	−0.09	−0.22*
Mistry	46	0.01	0.03
Moore	347	0.02	0.07
Morgan	357	−0.26**	−0.14**
Morris	412	−0.28**	−0.02
Murphy	248	0	0.03
O'Brien	110	−0.06	0.02
O'Neill	107	0.02	0.09
Patel	170	0.08	−0.05
Richardson	258	0.02	0.20**
Scott	348	0	0.26**
Shaw	232	−0.03	0.21**
Sullivan	94	0.03	−0.14**
Wright	500	0.11*	0.14**
Young	283	0.07	−0.08

[a]The number of cases varies, but n always equals 443, the number of districts.
[b]$*P < 0.05$; $**P < 0.01$.

surnames, a generally north–south orientation of the clines is more often seen. However, the polynomial regressions show that the clines are not strictly linear. Names high in the first two factors do not reach their highest relative frequencies in the far west (Cornwall) but at the longitude of Wales, and the south–north clines are also curved, with the highest frequencies generally being further north for surnames high in the second factor than for those especially high in the first.

Taking all the available evidence together, the general pattern of surname distributions in England and Wales is thus clear. There are no true isolates, but some isolation by distance is evident. Common surnames are polyphyletic and widely distributed, but not homogeneously. They form clines that account for statistically significant fractions of the geographic variation. Rare surnames are more localized and show steeper clines, but also more sampling variance than common names. Thus a surname analysis of local, regional and national samples supports the idea of clines rather than distinct subgroups, but many of the fine details must have been caused by specific events that will be hard or impossible to discover amidst the random sampling variation.

12 *Human population structure*

Human migration has been studied from many points of view. In using a surname model to study its effects, however, one is concerned with migration from a single angle: as the mechanism that redistributes genes geographically. Human migration draws pedigree lines on maps. The pattern of such lines depicts an aspect of human population structure of signficance to population genetics: it is the obverse of inbreeding. Such mapping of pedigree lines can be used to help explain distributions of human genetic polymorphisms and even, perhaps, to predict future redistributions – or, more exactly, to describe the conditions that would lead to alternative outcomes. Human genes cannot move except by the movements of people who carry them. At least that was true before the invention of artificial insemination. Therefore, historically, human migration accounted for all the movement of genes.

Gene movement may be seen in the distances from birthplaces of parents to the birthplaces of their children. A certain amount of tracing of individual pedigrees has been done by geneticists and others. Such studies inevitably have a geographic aspect. Pedigrees, however, are not representative of the population as a whole: they are more likely to be complete with respect to higher social classes, successful and noteworthy individuals and patients with hereditary diseases. Male ancestors are usually easier to identify and trace than female ones, so the male line is usually more complete than female and mixed lines in pedigrees. For these reasons a synthetic picture based on a collection of pedigrees is likely to be biased or to cover only the very few recent generations that can be completely ascertained.

An alternative approach to the structure of populations is through theoretical analyses. Some of these analyses are both elegant and general and can be applied to species with very diverse behaviours. A major branch of genetics (including human genetics) has been concerned with the mathematics of such analyses: the implications of population size for inbreeding, for instance. Problems sometimes arise in trying to apply the theory to concrete instances, however. It is not always clear how to measure demographic variables in such applications. In the theory, the size of the breeding population in the model is not the census count of the actual number of married people, but another number which (if the assumptions of the theory hold) would give the same result with respect to inbreeding as is found in the observed population. For instance, with the assumption of random breeding as part of the model, the size of the population in the model will be larger than the census count of parents if members of the society avoid incest. If the members of the society marry cousins by preference, however, the effect on size will be the reverse. If the question is not inbreeding but extinction of genes, the number may be a different one. When the task of the theorist is finished, it remains for the anthropologist to find the ratios of counts determined by censuses to those implicit in the theories and to discover the cultural as well as the biological factors that account for such differences.

Thus between the work of the Mendelian pedigree geneticist, on the one hand, and the theoretical population geneticist, on the other, there lies a level of analysis to be explored. It is the empirical study of the genetic structure of the human population. The purpose of such a study is to describe various samples of human populations in such a way that the population structure of the whole species can be characterized. The science is sometimes called genetic demography and it is through this science that anthropologists make one of their contributions to genetics. In doing so they study mating preferences and the way they work as well as simple census counts of the numbers of people in various categories.

Anthropologists have traditionally studied small isolated societies. The results of such studies may, to some extent, serve to model the general situation that obtained in the human population until recently. Since there is little migration into such societies, the situation may be relatively simple. Also the sizes of the populations may remain constant and thus another complication is avoided. However, the genetic demographer also needs to study larger societies. They are more common today and therefore more relevant to projections into the future. Also, only multiple studies of a number of small societies can possibly yield samples adequate for statistically significant generalization, but larger societies may more easily yield sample sizes suitable for this purpose.

One approach to the genetic structure of the whole species is through distribution studies of genetic markers. Thus Mourant and his colleagues (1976) have collected gene frequency data on human populations from all parts of the world. Distribution studies do permit the evaluation of the degree of geographic variability in human genetic polymorphisms such as blood groups. The fraction of variance within tribal populations can be compared with the fraction between tribal populations and much of the variation may be at the local level (Neel and Ward, 1970). However, direct application of such gene frequency data to the study of human population structure has limitations. In the first place, the immunological tests by which blood groups are determined are not free from methodological problems. Most studies of A_1, B and O blood groups can be compared directly without risk, but tests for some genetic polymorphisms are not always reliable. The next problem is sampling: blood donors with rare types may be over-represented and so may specific social or ethnic groups. Lastly, blood groups, transferrins, haptoglobins, haemoglobins, enzymes and other genetic products occur in non-human primates as well as in humans and, whatever their antiquity, polymorphisms trace their history over an indefinite period. Thus the effort to understand the processes of continuity and differentiation of subpopulations on the basis of polymorphisms is handicapped by the lack of an exact knowledge of the time when the differences found arose. Although it is possible to determine the blood group of a particular mummy, and sometimes of a dry bone, no large-scale population survey of any genetic polymorphism has been carried out on ancient material. It would be essentially impossible to do so with current methods. Studies of human population structure based on data concerning the geographic distribution of human polymorphisms can therefore tell little about process; on the contrary, such studies generally assume a process (migration) to explain the distributions found.

However, some investigators have combined world-wide studies of distributions of polymorphisms with theoretical models and computer simulation to show the plausibility of more complex interactions of processes. Thus Livingstone (1967, 1969) has plotted the geographic distributions of haemoglobin and thalassaemia polymorphisms and applied models involving selection as well as gene flow. The problem of lack of historical information remains, though, despite the interesting discussion and intepretation.

As the preceding chapters show, the gap in knowledge left by the shortcomings of other types of studies can be filled to some extent by analysis of the distribution of surnames. Despite all the difficulties which have been mentioned previously, surname distributions depict the result of migration and inbreeding over a sufficient length of time (in England about seven centuries) and of enough individuals (hundreds of thousands have been studied) to permit some generalizations about the genetic structure of the human population.

What is this structure of populations? It is sometimes thought of as consisting of all the factors that interrupt the genetic homogeneity of the species. Thus Morton (1982) defines structure as the 'factors that determine mating frequencies'. Structure is seen, then, in the non-random mating behaviour that divides species into genetically somewhat diverse descent lines and subpopulations.

For example, suppose one had a large hall the ceiling of which was covered with a map of England. Then the historical structure of the English population could be depicted by attaching strings at every spot where a person was resident at a certain time in the past, say the year AD 1400. A subsequent marriage or, more precisely, mating with viable offspring, could be represented by knotting the strings representing the two parents at a distance below the ceiling taken to represent the time somewhat subsequent to 1400 when the child was born and at a position directly underneath the place on the map where it was born. Further children of the same couple would be represented by tying together the two strings at lower points. Strings representing the children would begin at the knots of their birth and be knotted to the strings representing their mates at the positions representing the times and places of their children's births. If the height of the hall were equivalent to the time from 1400 to now, and on the floor there was an exact copy of the ceiling map, the present population of England would be represented by the strings that descended onto the floor map. The hall would be filled with a tangled macramé of strings and knots representing the complete pedigrees of the English population from AD 1400 to the present.

Of course there have been many immigrants to England in the intervening years. These could be represented by strings emerging from the walls at appropriate locations: few near the top because immigration was rare in the early years, but dense near the bottom because immigration (for instance of South Asians) has recently been heavy. This grading of the lateral strings from sparse above to dense below is an element of the structure. One way in which the strings can be seen as other than a random snarl is that on average they would be much more vertical near the ceiling than near the floor. In 1400, people were usually born, married and had their children in the same place or area. Near the floor, however, more of the strings would run nearly horizontally, as migration has become more frequent and the distances people move have become greater. This is shown very clearly in a study based on Church of England parish records of the Otmoor villages of Oxfordshire and Buckinghamshire during the last three centuries (Küchemann, Boyce and Harrison, 1967). Although the presence of long lateral strings is particularly characteristic of recent times, there were always some instances of long-distance migration, including some strings from the wall representing international migration. In some isolated regions of the world and in prehistoric times everywhere, these long strings represent intertribal matings and may have been rare. Long-distance filaments, however, are characteristic of the human species. No human group exists for very long without displaying this feature, although some may experience episodes of swamping by immigrants or isolation from external contacts for several generations in succession.

The pattern of strings will also show the effects of urbanization. In England many of the strings of all periods are directed towards London and other cities. One might think of the city as a magnet and of the strings as bits of steel wire. However, the analogy should not be carried too far: magnets attract all the wires within their range but cities attract only a fraction of the people because there are countervailing attractions of the countryside and suburbs and these exert different degrees of force on different individuals because of their social and economic statuses, previous experiences, learned attitudes and psychological dispositions. The same is true of general streams of population movement such as the southward trend of internal migration in Great Britain (Coleman, 1979, 1980*a*); usually only some of the strings run in the dominant direction. Mass movements, such as the migration of Irish to Liverpool during the potato famine in 1846 and 1847, are rare in human history.

Another feature of the breeding structure is the inbreeding loop. As one traces one's ancestors back in time, eventually some of the lines of descent reconverge on the same ancestor. This must inevitably be so because of the finite number of people. One has two parents, four grandparents, eight great grandparents and so on, doubling each generation to over a thousand in ten generations, over a million in twenty generations, over a thousand million in thirty, a million million in forty, and so on. There have never been anything approaching such numbers of people in the world, so before one reaches a pedigree of forty generations' depth there would have to be very dense networks of large inbreeding loops or somewhat less dense networks of more intimate inbreeding among close relatives.

Some theoretical analyses make a distinction: the extent of inbreeding implicit in the limited size of the population and consequent interrelatedness of all of its members produces a random component of the inbreeding coefficient – the amount of inbreeding that would occur if each individual had the same probability of mating with each member of the opposite sex. The extent by which the actual inbreeding exceeds this (because of tight inbreeding loops) or is exceeded by it (because of the avoidance of sexual relationships with close kin) is due to what is called the non-random component. In most societies the non-random component tends to be positive because tendencies to choose mates in the same vicinity or of the same social class, caste, religion or ethnic group make for partially distinct subpopulations within which matings take place more often than at random and within which interrelationships therefore accumulate with time. Rules of mating preference or prohibition vary greatly among societies. In some, incest tabus are extensive and strict and there is less inbreeding than expected at random. Nevertheless, in all societies there are at least occasional instances of small inbreeding loops and this is a significant aspect of the human population.

Inbreeding is cumulative. The inbreeding coefficient sums not only the loops leading to the individual, but also those which led to ancestors. This cumulative factor means that in the perspective of long periods of time the inbreeding coefficient approaches unity. This might not seem to be the case, because large inbreeding loops contribute very little to the coefficient. The formula for calculating inbreeding from a loop is

$$F = (\tfrac{1}{2})^{n-1},$$

in which F is the contribution of the loop to the inbreeding coefficient and n is the number of parent–child links involved in the loop. Technically a correction should be included:

$$F = (\tfrac{1}{2})^{n-1}(1 + F_A),$$

where F_A is the inbreeding coefficient of the common ancestor at the top of the loop. In practice F_A is usually unknown in human pedigrees. As an example of the small contribution of large loops, consider the loop for a person whose mother and father had a common ancestor 19 generations previously. The 20-generation loop would contain 40 links and its contribution to F would be $(\tfrac{1}{2})^{39}$, an amount so small as to be negligible even if there were thousands of such loops. As the number of generations increases, the number of antecedents doubles each generation and the chance of the same individual recurring in the pedigree is (at random) raised by something like 2 to the power of the number of generations. Meanwhile the loop that is formed has a contribution to inbreeding that decreases almost twice as fast ($\tfrac{1}{2}$ raised to the power of twice the number of generations less one).

Yasuda and Morton (1967) cited several studies in which the proportion of inbreeding from remote consanguinity (all relationships beyond first cousins once removed) ranged from 10 to 51% of the total. Societies in the upper part of this range with a high proportion of inbreeding from remote relationships are those with religious and/or legal restrictions against close intermarriage and little inbreeding of any kind. Thus one can safely say that, with the passage of time, the inbreeding coefficient approaches the asymptote of 1, not because of the inclusion of more and more remote common ancestors as much as because close inbreeding is cumulative generation after generation and, even if rare, will eventually be virtually complete. Since fewer generations are represented in small loops than in large ones, the small loops also tend to repeat themselves more often than the large ones. Data on small loops are also easier to collect than on large loops since people know much more about close than about distant relatives. For this reason, as well as the fact that the contribution of close inbreeding to the inbreeding coefficient generally represents the larger share, most of what is known about human inbreeding comes from the study of small loops over a few generations. Generalizations about the past from results of pedigree studies of inbreeding must therefore usually rest on the assumption that conditions have been stable and that recent experience is representative; to do so on the basis of analysis of surnames can incorporate data of the past, including information on the times of origin and historical frequency distributions.

As far as studies of the genetic structure of the British population are concerned, various methods have been applied. Roberts and Sunderland's (1973) compilation of studies of genetic variation in Britain brought together contributions by various authors on the history of migration, marital migration, distribution of blood groups and other genetically and partially genetically determined traits as well as special studies of such subpopulations as immigrant Sikhs, British gypsies, and the inhabitants of islands. Studies of isonymy were included (Dobson, 1973; Rawling, 1973), but most studies using the distribution of surnames to study the structure of the British population have been undertaken since Roberts and Sunderland's publication. As the previous pages demonstrate, surname studies, their shortcomings notwithstanding, now make a larger contribution to knowledge of the structure of the British and other populations than was the case even a few years ago.

In summary, to the extent that surnames are regularly inherited they can serve well in models of the genetic structure of populations. They have the advantage over gene products that: (1) large quantities of surname data are more easily available; (2) surnames yield more variants than polymorphic genes; (3) unlike genetic variants, the date of origin of surnames is known; and (4) unlike genes and their products, surnames can be studied on past populations. On the other hand common surnames often have multiple origins and rare ones usually do not yield enough information for statistically significant results. By combining information from various sources, however, surnames reveal much about the structure of human populations.

In areas such as Great Britain, the patterns of distribution of surnames in islands, rural villages, towns and cities are now known for hundreds of places and for the countries of England and Wales as a whole (see Appendix for examples of certain names). These studies of past and present surname distributions show that since the Middle Ages there has been an open population structure with considerable opportunity for the exchange of genes among geographic areas. In fact few studies anywhere suggest degrees of isolation sufficient to produce locally significant and enduring variations. Surname studies do suggest, however, that past inbreeding loops were more numerous than most attempts to count them imply. Hence there are fascicles of geographically localized descent lines held together by inbreeding, although sooner or later bound to other fascicles by migration and intermarriage.

Literature cited

Ammon, L. (1976) Smith and Jones: 1853 and 1975. *Population Trends*, **4**, 9–11.

Arner, G. B. L. (1908) *Consanguineous Marriages in the American Population*. Columbia University Studies in History, Economics and Public Law, vol. 31 no 3. Longman, Green and Co., New York.

Avcin, M. (1969) Gypsy isolates in Slovenia. *Journal of Biosocial Science*, **1**, 221–33.

Azevedo, E. S., daCosta, T. P., Silva, M. C. B. O. and Ribeiro, L. R. (1983) The use of surnames for interpreting gene frequency distribution and past racial admixture. *Human Biology*, **55**, 235–42.

Azevedo, E. S., Fortuna, C. M. M., Silva, K. M. C., Sousa, M. G. F., Machado, M. A., Lima, A. M. V. M. D. M., Aguiar, M. E., Abe, K., Eulalio, M. C. M. N., Conceicao, M. M., Silva, M. C. B. O. and Santos, M. G. (1982) Spread and diversity of human populations in Bahia, Brazil. *Human Biology*, **54**, 329–41.

Azevedo, E., Morton, N. E., Miki, C. and Yee, S. (1969) Distance and kinship in northeastern Brazil. *American Journal of Human Genetics*, **21**, 1–22.

Bailie, S. R. (1981) A surname analysis of two fishing communities in north-east Scotland. *Annals of Human Biology*, **8**, 392 (abstract).

Bailie, S. R. (1984) The structure of the population in fishing communities of north-east Scotland. PhD Dissertation, University of Aberdeen.

Bhalla, V. and Bhatia, K. (1976) Isonymy in a Bhatia leut. *Annals of Human Genetics*, **39**, 497–500.

Bhatia, K. and Wilson, S. R. (1981) The application of gene diversity analysis to surname diversity data. *Journal of Theoretical Biology*, **88**, 121–33.

Bourgoin, J. and Vu Tien Khang (1978) Quelques aspects de l'histoire génétique de quatre villages pyrenées depuis 1740. *Population*, **3**, 633–59.

Buchanan, A. V., Schwartz, R. J. and Weiss, K. M. (1982) *An Equivalence Class Name Library for Surnames from Vital Records in Laredo, Texas*. Second CDPG Report. Center for Demographic and Population Genetics, University of Texas Graduate School of Biomedical Sciences, Houston.

Buckatzsch, E. J. (1951) The constancy of local populations and migration in England before 1800. *Population Studies*, **8**, 62–9.

Cavalli-Sforza, L. L. and Bodmer, W. F. (1971) *The Genetics of Human Populations*, pp. 473–80. W. H. Freeman, San Francisco.

Chakraborty, R., Weiss, K. M., Rossman, D. L. and Norton, S. L. (1981) Distribution of last names: a stochastic model for likelihood determination in record linking. In *Genealogical Demography*, ed. B. Dyke and W. T. Morrill, pp. 63–9. Academic Press, New York.

Chen, Kuang-Ho and Cavalli-Sforza, L. L. (1983) Surnames in Taiwan: interpretation based on geography and history. *Human Biology*, **55**, 367–74.

Coleman, D. A. (1979) A study of migration and marriage in Reading, England. *Journal of Biosocial Science*, **11**, 369–89.

Coleman, D. A. (1980*a*) Some genetical inferences from the marriage system of Reading, Berkshire and its surrounding area. *Annals of Human Biology*, **7**, 55–76.

Coleman, D. A. (1980*b*) A note on the frequency of consanguineous marriages in Reading, England in 1972/1973. *Human Heredity*, **30**, 278–85.

Coleman, D. A. (1982) The population structure of an urban area in Britain. In *Current Developments in Anthropological Genetics*, vol. 2, ed. M. H. Crawford and J. H. Mielke, pp. 467–506. Plenum Press, New York.

Coleman, D. A. (1984) Marital choice and geographical mobility. In *Migration and Mobility*, ed. A. J. Boyce, Symposium of the Society for the Study of Human Biology 23, pp. 19–55. Taylor and Francis, London.

Crawford, M. H. (1980) The breakdown of reproductive isolation in an Alpine genetic isolate: Acceglio, Italy. In *Population Structure and Genetic Disorders*, ed. A. W. Eriksson, H. R. Forsius, H. R. Nevanlinna, P. L. Workman and R. K. Norio, pp. 57–71. Academic Press, New York and London.

Crow, J. F. (1980) The estimation of inbreeding from isonymy. *Human Biology*, **52**, 1–14.

Crow, J. F. (1983) Discussion. In: Surnames as markers of inbreeding and migration (arranged K. Gottlieb). *Human Biology*, **55**, 383–97.

Crow, J. F. and Mange, A. P. (1965) Measurement of inbreeding from the frequency of marriages between persons of the same surname. *Eugenics Quarterly*, **12**, 199–203.

Darwin, G. H. (1875) Marriages between first cousins in England and their effects. *Journal of the Statistical Society*, **38**, 153–84.

Devor, E. J. (1980) Marital structure and genetic isolation in a rural Hispanic population in northern New Mexico. *American Journal of Physical Anthropology*, **53**, 257–65.

Devor, E. J. (1983) Matrix methods for the analysis of isonymous and nonisonymous surname pairs. *Human Biology*, **55**, 277–88.

Devor, E. J., Crawford, M. H. and Koertvelyessy, T. (1984) Marital structure and genetic heterogeneity of Ramea Island, Newfoundland. *American Journal of Physical Anthropology*, in press.

Dobson, T. (1973) Historical population structure in Northumberland. In *Genetic Variation in Britain*, ed. D. F. Roberts and E. Sunderland, pp. 67–81. Taylor and Francis, London.

Dobson, T. and Roberts, D. F. (1971) Historical population movement and gene flow in Northumberland parishes. *Journal of Biosocial Science*, **3**, 191–208.

Dyke, B., James, A. V. and Morrill, W. T. (1983) Estimation of random isonymy. *Annals of Human Biology*, **10**, 295–8.

Ellis, W. S. and Friedl, J. (1976) Inbreeding as measured by isonymy and by pedigrees in Kippel, Switzerland. *Social Biology*, **23**, 158–67.

Ellis, W. S. and Starmer, W. T. (1978) Inbreeding as measured by isonymy, pedigrees, and population size in Törbel, Switzerland. *American Journal of Human Genetics*, **30**, 366–76.

Floris, G. and Vona, G. (1980) Isonomie maritale et coefficient de parenté entre six communes de l'île de Sardaigne. *L'Anthropologie*, **84**, 300–6.
Fox, W. R. and Lasker, G. W. (1983) The distribution of surname frequencies. *International Statistical Review*, **51**, 81–7.
Friedl, J. and Ellis, W. S. (1974) Inbreeding, isonymy, and isolation in a Swiss community. *Human Biology*, **46**, 699–712.
Galton, F. (1889) Probable extinction of families. In *Natural Inheritance*, appendix F, pp. 241–48. Macmillan, New York.
Gottlieb, K. (1983) Genetic demography of Denver, Colorado: Spanish surname as a marker of Mexican ancestry. *Human Biology*, **55**, 227–34.
Guppy, H. B. (1890) *Homes of Family Names of Great Britain*. Harrison and Sons, London.
Halberstein, R. A. and Crawford, M. H. (1975) Demographic structure of a transplanted Tlaxcalan population in the Valley of Mexico. *Human Biology*, **47**, 201–32.
Hiorns, R. W., Harrison, G. A. and Küchemann, C. F. (1973) Factors affecting the genetic structure of populations: an urban–rural contrast in Britain. In *Genetic Variation in Britain*, ed. D. F. Roberts and E. Sunderland, pp. 17–32. Taylor and Francis, London.
Hiorns, R. W., Harrison, G. A., Boyce, A. J. and Küchemann, C. F. (1969) A mathematical analysis of the effect of movement on the relatedness between populations. *Annals of Human Genetics*, **32**, 237–50.
Hurd, J. P. (1983) Comparison of isonymy and pedigree analysis measures in estimating relationships between three 'Nebraska' Amish churches in central Pennsylvania. *Human Biology*, **55**, 349–55.
Hussels, I. (1969) Genetic structure of Saas, a Swiss isolate. *Human Biology*, **41**, 469–79.
Imaizumi, Y. (1974) Genetic structure in the United Kingdom. *Human Heredity*, **24**, 151–9.
Jacquard, A. (1975) Inbreeding: one word several meanings. *Theoretical Population Biology*, **7**, 338–63.
James, A. V. (1983) Isonymy and mate choice on St. Bart, French Indies: computer simulations of random and total isonymy. *Human Biology*, **55**, 297–303.
Kamizaki, M. (1954) Frequency of isonymous marriages. *Seibutsu Tokei-gaku Zassi*, **2**, 292–98. (In Japanese. Cited after Yasuda, 1983.)
Kaplan, B. A. and Lasker, G. W. (1983) The present distribution of some English surnames derived from place names. *Human Biology*, **55**, 243–50.
Kaplan, B. A., Lasker, G. W. and Chiarelli, B. (1978) Communality of surnames: a measure of biological inter-relationships among 31 settlements in upper Val Varieta in the Italian Alps. *American Journal of Physical Anthropology*, **49**, 251–6.
Kashyap, L. K. (1980) Trends of isonymy and inbreeding among the Ahamdiyyas of Kashmir. *Journal of Biosocial Science*, **12**, 219–25.
Kashyap, L. K. and Tiwari, S. C. (1980) Kinetics of genetic kinship as inferred by isonymy among Ahmadiyyas of Kashmir Valley. *Human Biology*, **52**, 311–24.

Katayama, K. and Toyomasu, T. (1979) A genetic study on the local populations in Mie Prefecture. IV. The genetic relations among the Kamishimp, Toshi, Momotori and Toba populations. *Journal of the Anthropological Society of Nippon*, **87**, 377–92.

Katayama, K., Toyomasu, T. and Matsumoto, H. (1978) Genetic study on the local populations in Mie Prefecture. II. Population structure in Kamishima Island. *Journal of the Anthropological Society of Nippon*, **86**, 83–94.

Kirkland, J. R. and Jantz, R. L. (1977) Inbreeding, marital movement, and genetic isolation of a rural Appalachian population. *Annals of Human Biology*, **4**, 211–18.

Koertvelyessy, T., Crawford, M. H. and Huntsman, R. G. (1984) The analysis of matings in selected Newfoundland communities with isonymy and surname-pair-matrices. *American Journal of Physical Anthropology*, **63**, 179 (abstract).

Kosten, M., Williams, J. and Mitchell, R. J. (1983) Historical population structure of two Tasmanian communities using surname analysis. *Journal of Biosocial Science*, **15**, 367–76.

Küchemann, C. F., Boyce, A. J. and Harrison, G. A. (1967) A demographic and genetic study of a group of Oxfordshire villages. *Human Biology*, **39**, 251–76.

Küchemann, C. F., Lasker, G. W. and Smith, D. I. (1979) Historical changes in the coefficient of relationship by isonymy among the populations of the Otmoor villages. *Human Biology*, **51**, 63–77.

Lalouel, J. M. and Langaney, A. (1976) Bedik and Niokholonko of Senegal: intervillage relationship inferred from migration data. *American Journal of Physical Anthropology*, **45**, 453–66.

Lasker, G. W. (1968) The occurrence of identical (isonymous) surnames in various relationships in pedigrees: a preliminary analysis of the relation of surname combinations to inbreeding. *American Journal of Human Genetics*, **20**, 250–7.

Lasker, G. W. (1969) Isonymy (occurrence of the same surname in affinal relatives): a comparison of rates calculated from pedigrees, grave markers and death and birth registers. *Human Biology*, **41**, 309–21.

Lasker, G. W. (1977) A coefficient of relationship by isonymy: a method for estimating the genetic relationship between populations. *Human Biology*, **49**, 489–93.

Lasker, G. W. (1978*a*) Relationships among the Otmoor villages and surrounding communities as inferred from surnames contained in the current register of electors. *Annals of Human Biology*, **5**, 105–11.

Lasker, G. W. (1978*b*) Increments through migration to the coefficient of relationship between communities by isonymy. *Human Biology*, **50**, 235–40.

Lasker, G. W. (1983) The frequencies of surnames in England and Wales. *Human Biology*, **55**, 331–40.

Lasker, G. W., Chiarelli, B., Masali, M., Fedele, F. and Kaplan, B. A. (1972) Degree of human genetic isolation measured by isonymy and marital distances in two communities in an Italian Alpine valley. *Human Biology*, **44**, 351–60.

Lasker, G. W., Coleman, D. A., Aldridge, N. and Fox, W. R. (1979) Ancestral relationships within and between districts in the region of Reading, England, as estimated by isonymy. *Human Biology*, **51**, 445–60.

Lasker, G. W. and Kaplan, B. A. (1983) English place-name surnames tend to cluster near the place named. *Names*, **31**, 167–77.

Lasker, G. W., Kaplan, B. A. and Mascie-Taylor, C. G. N. (1984) And who is thy neighbor? Biological relationship in the English village. *Humanbiologia Budapestinensis*, in press.

Lasker, G. W. and Mascie-Taylor, C. G. N. (1983) Surnames in five English villages: relationship to each other, to surrounding areas, and to England and Wales. *Journal of Biosocial Science*, **15**, 25–34.

Lasker, G. W. and Roberts, D. F. (1982) Secular trends in relationship as estimated by surnames: a study of a Tyneside parish. *Annals of Human Biology*, **9**, 299–307.

Lasker, G. W., Wetherington, R. K., Kaplan, B. A. and Kemper, R. V. (1983) Isonymy between two towns in Michoacán. In *Estudios de Antropología Biologica*, pp. 159–63. Universidad Nacional Autonoma de Mexico.

Lasker, M. (1941) Essential fructosuria. *Human Biology*, **13**, 51–63.

Lewitter, F. I., Hurwich, B. J. and Nubani, N. (1983) Tracing kinship through father's first name in Abu Ghosh, an Israeli Arab patrilineal society. *Human Biology*, **55**, 375–81.

Livingstone, F. B. (1967) *Abnormal Hemoglobins in Human Populations*. Aldine, Chicago.

Livingstone, F. B. (1969) Gene frequency clines of the beta hemoglobin locus in various human populations and their simulation by models involving differential selection. *Human Biology*, **41**, 223–36.

Lotka, A. J. (1931) The extinction of families. *Journal of the Washington Academy of Science*, **21**, 377–453.

McClure, P. (1979) Patterns of migration in the Late Middle Ages: the evidence of English place-name surnames. *Economic History Review* (2nd series), **32**, 167–82.

McKinley, R. A. (1976) The distribution of surnames derived from the names of some Yorkshire towns. In *Essays Presented to Marc Fitch by Some of his Friends*, ed. F. G. Emmison and R. Stephens. Leopard's Head Press, London.

Malécot, G. (1948) *Les mathématiques de l'hérédité*. Masson, Paris.

Mascie-Taylor, C. G. N. and Lasker, G. W. (1984) Geographic distribution of surnames in Britain: the Smiths and Joneses have clines like blood group genes. *Journal of Biosocial Science*, **16**, 301–8.

Morton, N. E. (1972) Pingelap and Mokil Atolls: clans and cognate frequencies. *American Journal of Human Genetics*, **24**, 290–8.

Morton, N. E. (1973) *Genetic Structure of Populations*. University of Hawaii, Honolulu.

Morton, N. E. (1982) *Outline of Genetic Epidemiology*. Karger, New York.

Morton, N. E. and Hussels, I. (1970) Demography of inbreeding in Switzerland. *Human Biology*, **42**, 65–78.

Morton, N. E., Klein, D., Hussels, I., Donival, P., Todorov, A., Lew, R. and Yee, S. (1973) Genetic structure of Switzerland. *American Journal of Human Genetics*, **25**, 347–61.

Morton, N. E., Smith, C., Hill, R., Frankiewicz, A., Law, P. and Yee, S. (1976) Population structure of Barra (Outer Hebrides). *Annals of Human Genetics*, **39**, 339–52.

Morton, N. E., Yee, S., Harris, D. E. and Lew, R. (1971) Bioassay of kinship. *Theoretical Population Biology*, **2**, 507–24.

Mourant, A. E., Kopeć, A. C. and Domaniewska-Sobczak, K. (1976) The *Distribution of the Human Blood Groups and other Polymorphisms*, 2nd edn. Oxford University Press, London.

Neel, J. V. and Ward, R. H. (1970) Village and tribal genetic distances among American Indians, and possible implications for human evolution. *Proceedings of the National Academy of Sciences, USA*, **65**, 323–30.

Pollitzer, W. S. and Brown, W. H. (1969) Survey of demography, anthropometry, and genetics in the Melungeons of Tennessee: an isolate of hybrid origin in process of dissolution. *Human Biology*, **41**, 388–400.

Raspe, P. D. (1983) The mating structure of a subdivided population. PhD Dissertation, University of Cambridge.

Raspe, P. and Lasker, G. W. (1980) Biological relationships by isonymy among the Scilly Isles and of the Scilly Isles with the rest of England and Wales. *Annals of Human Biology*, **7**, 401–10.

Rawling, C. P. (1973) A study of isonymy. In *Genetic Variation in Britain*, ed. D. F. Roberts and E. Sunderland, pp. 83–93. Taylor and Francis, London.

Reaney, P. H. (1976) *A Dictionary of British surnames*, 2nd edn. Routledge and Kegan Paul, London.

Refshauge, W. F. and Walsh, R. J. (1981) Pitcairn Island: fertility and population growth, 1790–1856. *Annals of Human Biology*, **8**, 303–12.

Registrar General (1853) *Annual Report*, pp. xvii–xxviii. HMSO, London.

Reid, R. M. (1973) Inbreeding in human populations. In *Methods and Theories of Anthropological Genetics*, ed. M. H. Crawford and P. L. Workman, pp. 83–116. University of New Mexico Press, Albuquerque.

Relethford, J. H. (1982) Isonymy and population structure of Irish isolates during the 1890s. *Journal of Biosocial Science*, **13**, 317–36.

Relethford, J. H. and Lees, F. C. (1983) Correlation analysis of distance measures based on geography, anthropometry and isonymy. *Human Biology*, **55**, 653–65.

Relethford, J. H., Lees, F. C. and Crawford, M. H. (1981) Population structure and anthropometric variation in rural western Ireland: isolation by distance and analysis of the residuals. *American Journal of Physical Anthropology*, **55**, 233–45.

Roberts, D. F. (1980) Inbreeding and ecological change: an isonymic analysis in a Tyneside parish over three centuries. *Social Biology*, **27**, 230–40.

Roberts, D. F. (1982) Population structure of farming communities of northern England. In *Current Developments in Anthropological Genetics*, vol. 2, ed. M. H. Crawford and J. H. Mielke, pp. 367–84. Plenum Press, New York.

Roberts, D. F. and Rawling, C. P. (1974) Secular trends in genetic structure: an isonymic analysis of Northumberland parish records. *Annals of Human Biology*, **1**, 393–410.

Roberts, D. F. and Roberts, M. J. (1983) Surnames and relationships: an Orkney study. *Human Biology*, **55**, 341–7.

Roberts, D. F. and Sunderland, E. (1973) *Genetic Variation in Britain*. Symposia of the Society for the Study of Human Biology, vol. 12. Taylor and Francis, London.

Robinson, A. P. (1983) Inbreeding as measured by dispensations and isonymy on a small Hebridean island, Eriksay. *Human Biology*, **55**, 289–95.

Robinson, A. P. and Clegg, E. J. (1981) Aspects of population structure in Eriksay. Society for the Study of Human Biology, Abstracts of Communications, p. 7. (No place of publication listed.)

Rogers, L. A. (1984) Phylogenetic identification of a religious isolate and the measurement of inbreeding. PhD Dissertation, University of Kansas, Lawrence.

Schull, W. J. and Neel, J. V. (1965) *The Effects of Inbreeding on Japanese Children*. Harper and Row, New York.

Shaw, R. F. (1960) An index of consanguinity based on the use of the surname in Spanish-speaking countries. *Journal of Heredity*, **51**, 221–30.

Simmons, R. T., Graydon, J. J. and Tindale, N. B. (1964) Further blood group genetical studies of Australian Aborigines of Bentinck, Mornington and Forsyth Islands and the mainland Gulf of Carpentaria, together with frequencies for natives of the Western Desert, Western Australia. *Oceania*, **35**, 66–80.

Simmons, R. T., Tindale, N. B. and Birdsell, J. B. (1962) A blood group genetical survey in Australian Aborigines of Bentinck, Mornington and Forsyth Islands, Gulf of Carpentaria. *American Journal of Physical Anthropology*, **20**, 303–20.

Smith, M. T. and Hudson, B. L. (1984) Isonymic relationships in the parish of Fylingdales, North Yorkshire in 1851. *Annals of Human Biology*, **11**, 141–8.

Smith, M. T., Smith, B. L. and Williams, W. R. (1984) Changing isonymic relationships in Fylingdales Parish, North Yorkshire, 1841–1881. *Annals of Human Biology*, in press.

Sorg, M. H. (1983) Isonymy and diabetes prevalence in the island population of Vinalhaven, Maine. *Human Biology*, **55**, 303–11.

Souden, D. and Lasker, G. W. (1978) Biological inter-relationships between parishes in East Kent: an analysis of Marriage Duty Act returns for 1705. *Local Population Studies*, **21**, 30–9.

Stevenson, J. C., Brown, R. J. and Schanfield, S. M. (1983) Surname analysis as a sampling method for recovering genetic information. *Human Biology*, **55**, 219–25.

Sunderland, E. (1982) The population structure of the Romany Gypsies. In *Current Developments in Anthropological Genetics*, vol. 2, ed. M. H. Crawford and J. H. Mielke, pp. 125–37. Plenum Press, New York.

Swedlund, A. C. (1971) The genetic structure of an historical population: a study of marriage and fertility in Deerfield, Massachusetts. Research Reports in Anthropology No. 7. University of Massachusetts, Amherst.

Swedlund, A. C. (1975) Isonymy: estimating inbreeding from social data. *Bulletin of the Eugenics Society*, **7**, 67–73.

Swedlund, A. C. and Boyce, A. J. (1983) Mating structure in historical populations: estimation by analysis of surnames. *Human Biology*, **55**, 251–62.

Ward, R. H. and Neel, J. V. (1970) Gene frequencies and microdifferentiation among Makiritari Indians. IV. A comparison of a genetic network with ethnohistory and migration matrices, a new index of genetic isolation. *American Journal of Human Genetics*, **22**, 538–61.

Watson, R. (1975) A study of surname distribution in a group of Cambridgeshire parishes, 1538–1840. *Local Population Studies*, **15**, 23–32.

Weiss, K. M. and Chakraborty, R. (1982) Genes, populations, and disease, 1930–1980: a problem-oriented review. In *A History of American Physical Anthropology 1930–1980*, ed. F. Spencer, pp. 371–404. Academic Press, New York.

Weiss, K. M., Chakraborty, R., Buchanan, A. V. and Schwartz, R. J. (1983) Mutations in names: implications for assessing identity by descent from historical records. *Human Biology*, **55**, 313–22.

Weiss, V. (1973) Eine neue Methode zur Schätzung des Inzuchtkoeffizienten aus den Familiennamenhäufigkeiten der Vorfahren. *Biologische Rundschau*, **11**, 314–15.

Weiss, V. (1974) Die Verwendung von Familiennamenhäufigkeiten zur Schätzung der genetischen Verwandtschaft. Ein Beitrag zur Populationsgenetik des Vogtlandes. *Ethnografische-Archäologische Zeitschrift*, **15**, 433–51.

Wijsman, E., Zei, G., Moroni, A. and Cavalli-Sforza, L. L. (1984) Surnames in Sardinia. II. Computation of migration matrices from surname distributions in different periods. *Annals of Human Genetics*, **48**, 65–78.

Wilson, S. R. (1981) The analysis of g-isonymy data. *Annals of Human Biology*, **8**, 341–50.

Wright, S. (1922) Coefficients of inbreeding and relationship. *American Naturalist*, **56**, 330–8.

Yasuda, N. (1983) Studies of isonymy and inbreeding in Japan. *Human Biology*, **55**, 263–76.

Yasuda, N., Cavalli-Sforza, L. L., Skolnick, M. and Moroni, A. (1974) The evolution of surnames: and analysis of their distribution and extinction. *Theoretical Population Biology*, **5**, 123–42.

Yasuda, N. and Furusho, T. (1971*a*) Random and non-random inbreeding revealed from isonymy study. I. Small cities of Japan. *American Journal of Human Genetics*, **23**, 303–16.

Yasuda, N. and Furusho, T. (1971*b*) Random and non-random inbreeding revealed from isonymy study. II. A group of farm villages in Japan. *Japanese Journal of Human Genetics*, **15**, 231–40.

Yasuda, N. and Morton, N. E. (1967) Studies in human population structure. In *Proceedings of the Third International Congress of Human Genetics*, ed. J. F. Crow and J. V. Neel, pp. 249–65. Johns Hopkins University Press, Baltimore.

Yasuda, N. and Saitou, N. (1984) Random isonymy and surname distribution in Japan. *Biology and Society*, **1**, 75–84.

Zei, G., Guglielmino, C. R., Siri, E., Moroni, A. and Cavalli-Sforza, L. L. (1983*a*) Surnames as neutral alleles: observations in Sardinia. *Human Biology*, **55**, 357–65.

Zei, G., Guglielmino Matessi, R., Siri, E., Moroni, A. and Cavalli-Sforza, L. L. (1983*b*) Surnames in Sardinia. I. Fit of frequency distributions for neutral alleles and genetic population structure. *Annals of Human Genetics*, **47**, 329–52.

Zipf, G. K. (1949) *Human Behavior and the Principle of Least Effort*. Addison-Wesley, Reading, Mass.

Appendix: maps and diagrams of 100 surnames in England and Wales

Prepared by C. G. N. Mascie-Taylor*, A. J. Boyce† and G. Brush*

Past studies permitting the mapping of surnames in Britain are limited. Guppy (1890) published data on the distribution in Great Britain of the surnames of land-owning farmers, whom he considered to be the geographically most stable element in the society. Unfortunately Guppy only recorded part of the data – the frequency of surnames in the counties in which they reached or exceeded 7 per thousand; thus data on even the most common surnames were lacking for some counties.

Through the cooperation of the Office of Population Censuses and Surveys of the Registrar General's office, it has been possible to acquire, edit and computerize an alphabetically arranged list of the number of persons of each of a selected list of surnames married in each registration district of England and Wales in the first three months of 1975. Some analyses of these data have been undertaken (Lasker, 1983; Mascie-Taylor and Lasker, 1984; G. W. Lasker and C. G. N. Mascie-Taylor, in preparation). We here add maps and diagrams of the geographic distributions of these surnames.

Marriage records are generally preferable for distribution studies to birth or death records because the population sampled by marriage records is the adult breeding population of interest in human population genetics, whereas some individuals listed in birth and death records never lived to enter the breeding population. Although most individuals listed in the marriage records were resident in the district where the marriage was registered, there are some exceptions, especially among the bridegrooms. The difference between the two sexes in this regard is apparently small, however, because there is not a significantly greater tendency for a surname to be more localized among females that bear it than among males (Lasker, 1983).

* Department of Physical Anthropology, University of Cambridge.
† Department of Biological Anthropology, University of Oxford.

Maps

In the maps, surname frequencies of 1–4 per district are indicated by a dot (.) at the centre of population of the district in which they occur; small rectangles (▪) represent frequencies of 5–9 individuals, and larger rectangles (■), frequencies of 10 or more.

Graphs

The graphs depict, for each surname, the probabilities of local excesses or deficiencies in frequency from west to east and from south to north. The number of occurrences of each surname expected in each district if all surnames had a uniform geographic distribution was subtracted from the observed number of occurrences and a measure of the probability of each deviation occurring by chance was recorded. The west–east and south–north distributions of these values were fitted by the curves shown on the right-hand pages. The degree of departure of points on these curves from zero is thus an indication of the probability of increased (or decreased) frequency of the surname at the longitude or latitude at the number of kilometres east (left-hand diagram) or north (right-hand diagram) of the Ordnance reference point.

ADAMS: A linear cline in frequencies descends toward the north. According to Guppy, *Adams* was rare in the eastern and northern counties but well represented in the counties on the Welsh border.

ALLEN: The frequency increases somewhat towards the east. In the past, the principal centre seems to have been in the Midlands, but the surname was rare in the north except the extreme north-east.

ANDERSON: Now common in Northumberland, in Guppy's time there were 100 per 10000 in Scotland and 74 per 10000 in Northumberland, the northernmost county of England.

BAILEY: There are no significant clines. 'With the exception of the northern counties of England and of the four south-western counties, its distribution is pretty general' (Guppy, pp. 24–5).

For explanation of maps and graphs see p. 91

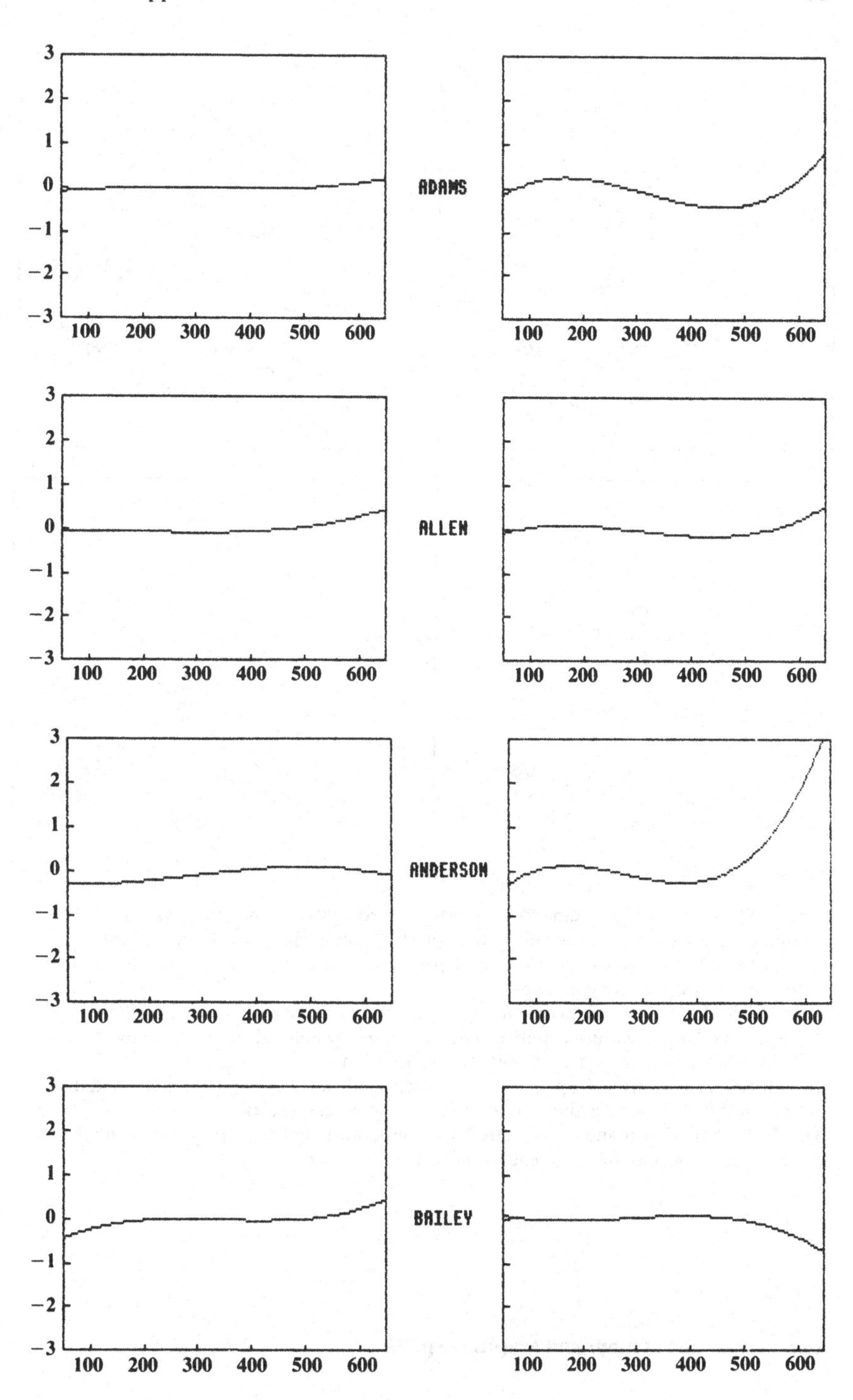
3
2
1
0
−1
−2
−3
100 200 300 400 500 600
ADAMS
ALLEN
ANDERSON
BAILEY

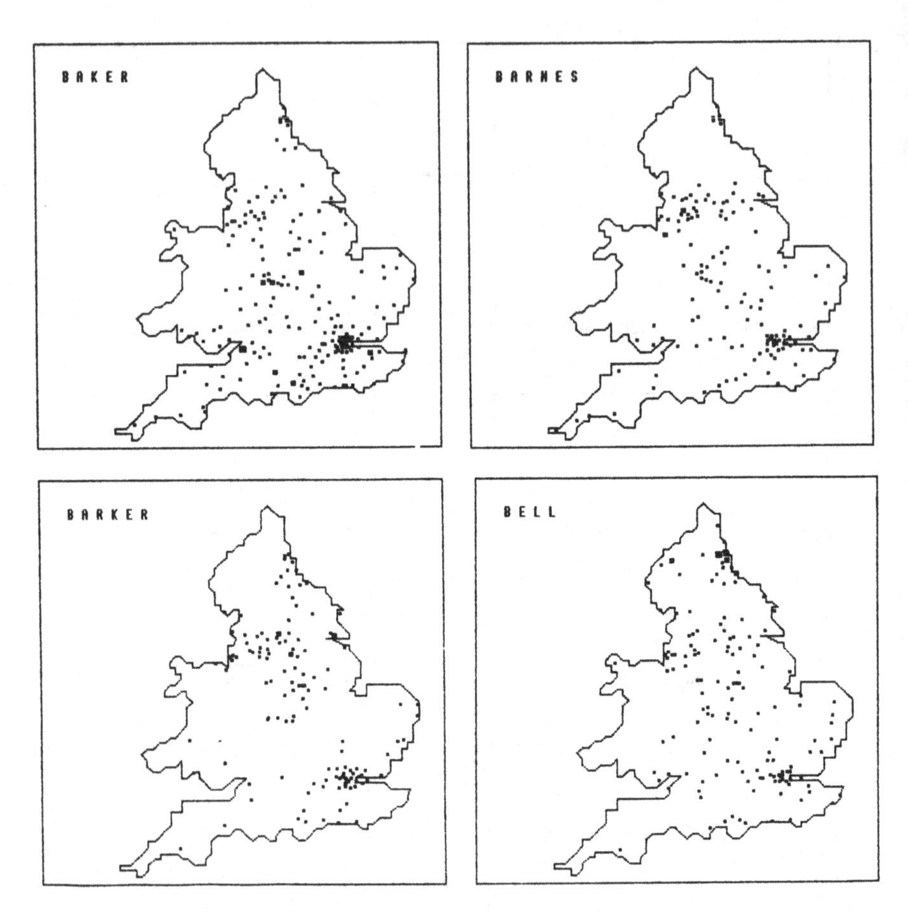

BAKER: Has a cline ascending to the south. According to Guppy, *Baker* was most numerous in the south of England and diminished rapidly in frequency northward; it abounded in the coastal counties of southern England (except Cornwall and Devon, the two most south-western counties).

BARKER: In the nineteenth century it was said to be confined to the northern half of England and to the eastern counties, whereas *Tanner* was its substitute in the south. *Barker* still has a predominantly northern distribution.

BARNES: According to Guppy, this had two principal homes, one in the south and the other in the north of England. There are now no appreciable clines.

BELL: Was previously and still is centred near the Scottish border where persons with the name were reported to be at home by the sixteenth century.

For explanation of maps and graphs see p. 91

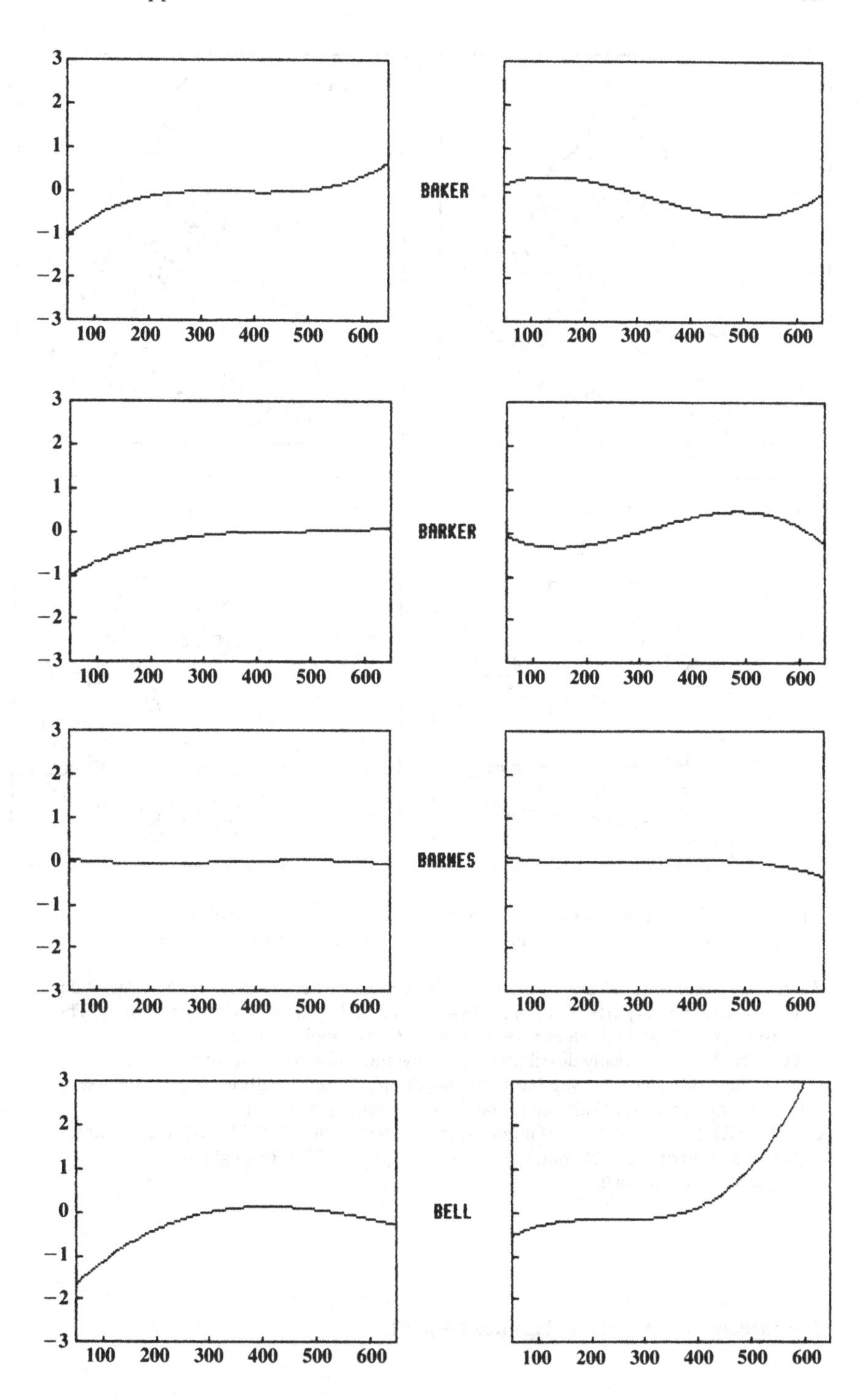
BAKER
BARKER
BARNES
BELL
3
2
1
0
−1
−2
−3
100
200
300
400
500
600

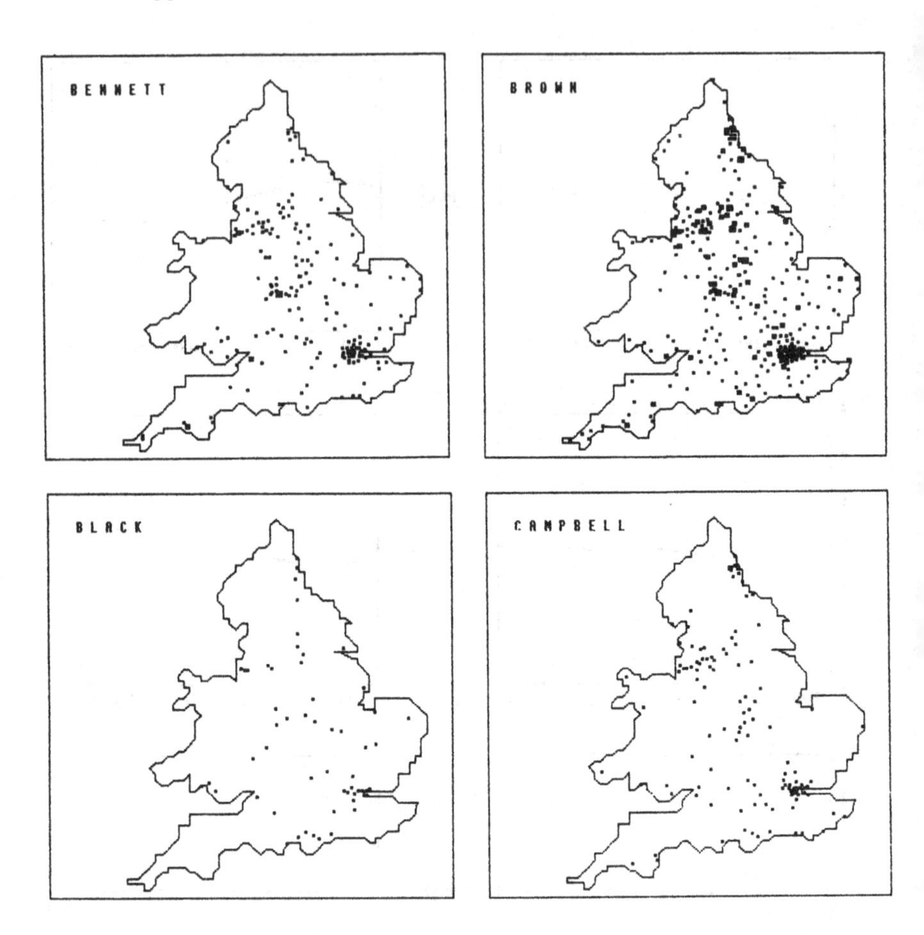

BENNETT: According to Guppy it was rare or absent north of Lincolnshire but well dispersed over the rest of England. There are now no strong directional trends in frequencies.
BLACK: Occurred in 35 per 10000 people in Scotland, where it was fairly general, and it was also common in parts of the far north and Midland England, according to Guppy. The sample for 1975 is too small to reveal significant geographic variation.
BROWN: Was universally distributed over England and a large part of Scotland, according to Guppy; however, it was rare bordering, and especially in, Wales. At present it is general except for Wales and especially common in the north.
CAMPBELL: 'One of the most numerous and powerful of the Highland clans, and under the leadership of the noble house of Argyll' (Guppy, p. 597). In England we find it most frequent in the far north.

For explanation of maps and graphs see p. 91

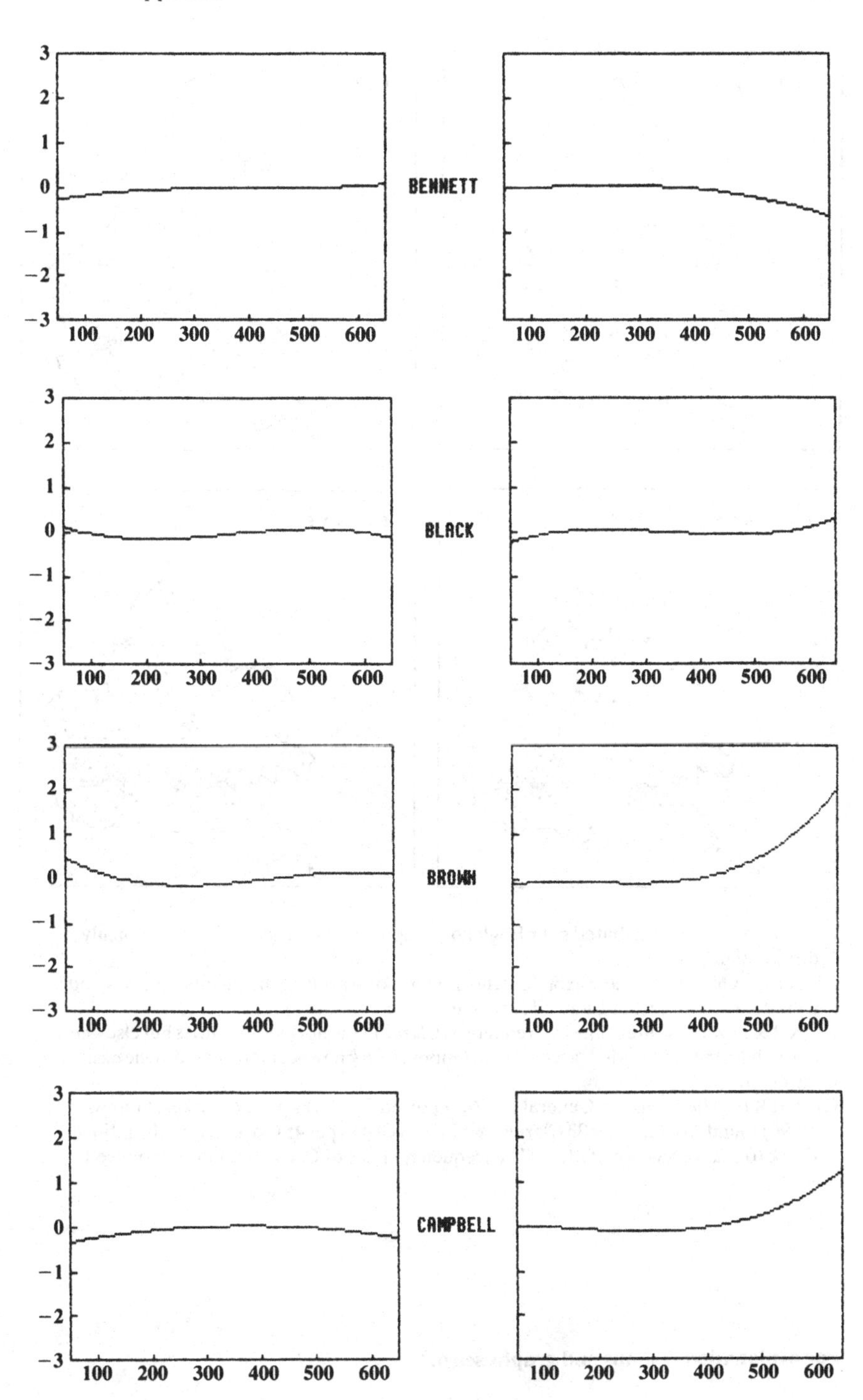

3
2
1
0
−1
−2
−3
100
200
300
400
500
600
BENNETT
BLACK
BROWN
CAMPBELL

CARTER: 'Well distributed over England' (Guppy, p. 28). Today it is not regionally differentiated.

CHAPMAN: The Anglo-Saxon for 'cheap-men', or travelling merchants, this was and is a predominantly east of England surname.

CLARK: Twice or three times as frequent as *Clarke* in the northern counties but elsewhere in England the ratio varied, according to Guppy. *Clark* now seems focussed in the east and far north.

CLARKE: The Registrar General (1856) reported 38 Clarks per 29 Clarkes; Guppy (1890) found 38 Clarks per 33 Clarkes; we find 38 Clarks per 45 Clarkes: thus the ratio of *Clark* to *Clarke* has been falling. The frequency clines of Clarke are not pronounced.

For explanation of maps and graphs see p. 91

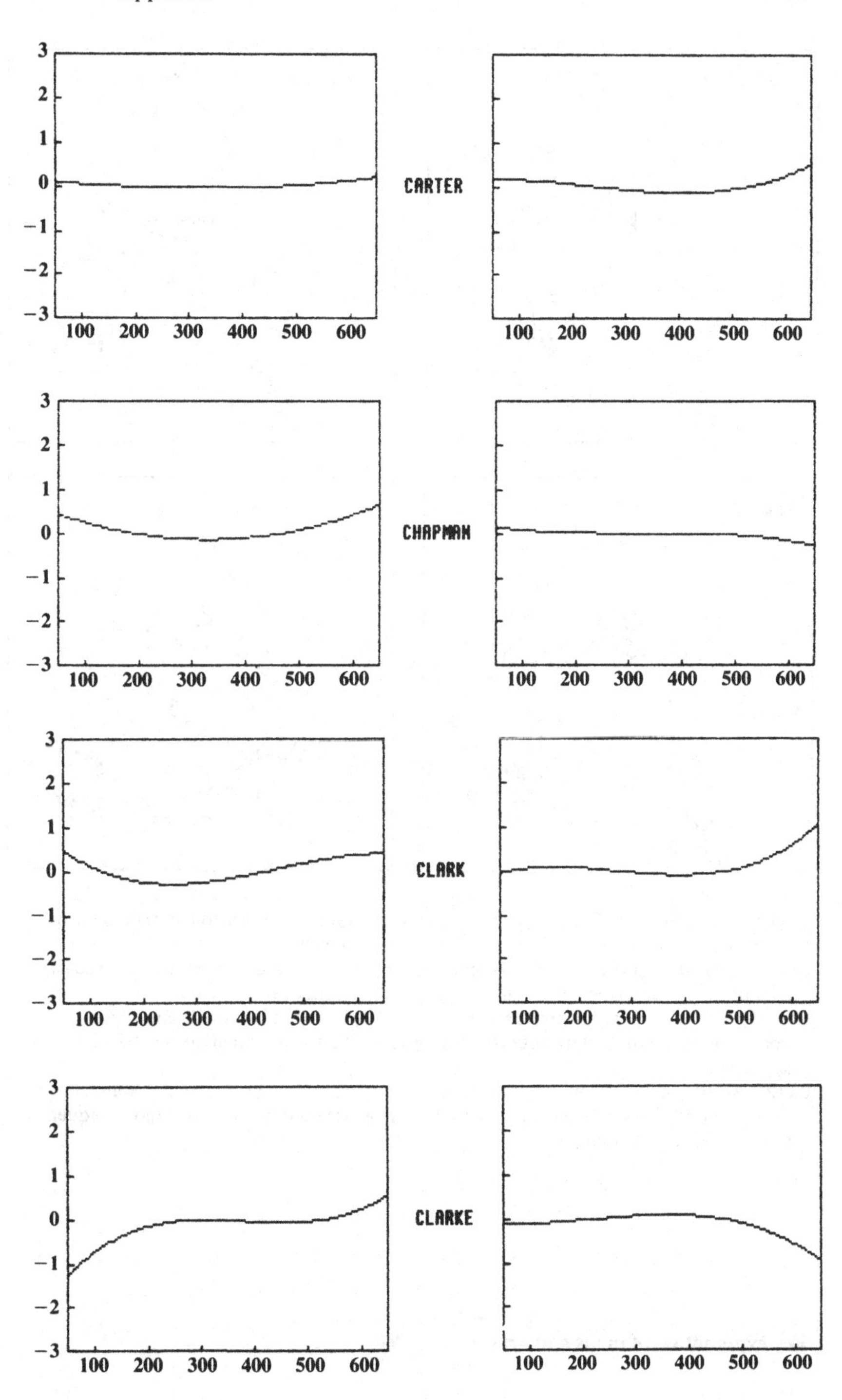
3
2
1
0
−1
−2
−3
100 200 300 400 500 600
CARTER
CHAPMAN
CLARK
CLARKE

COLLINS: A south of England name, according to Guppy, who also found it common in South Wales. It now appears most frequently in Cornwall.
COOK: 'Most frequent in the south-central counties of England and in the eastern coast counties . . .' (Guppy. p. 29). *Cook* now rises in frequency towards the east.
COOPER: 'Distributed over the greater part of England, but rare or absent in the northern and south-western counties' (Guppy, p. 29). The dearth in the north is still apparent.
COX: According to Guppy, rare or absent in the north of England and in the counties of the east coast. These trends are not now apparent and the name seems to be more frequent towards the east and south.

For explanation of maps and graphs see p. 91

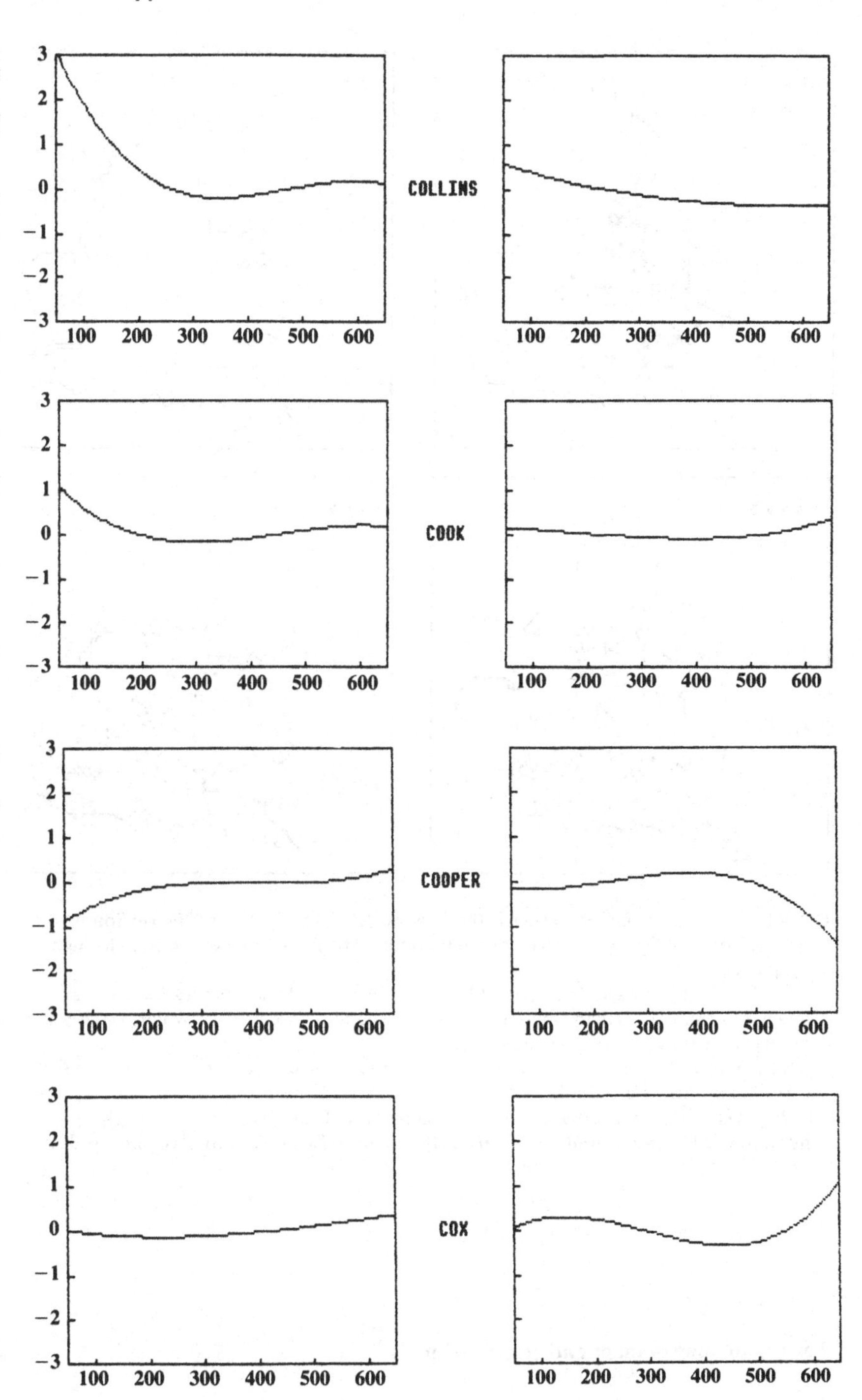

3
2
1
0
−1
−2
−3
100
200
300
400
500
600
COLLINS
COOK
COOPER
COX

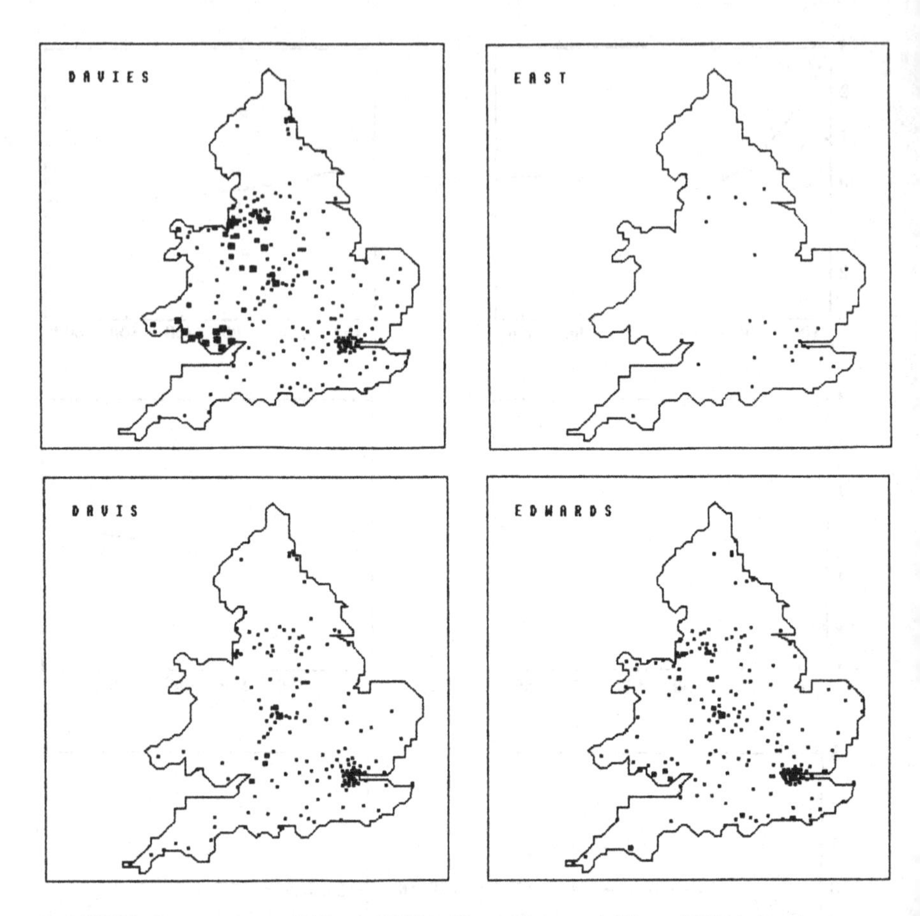

DAVIES: Guppy reported 500 per 10000 in North Wales and 600 per 10000 in South Wales. Today it displays a distribution typical for a Welsh surname with especially high frequencies in South Wales.

DAVIS: Is sparse towards the north of England. Guppy said that whereas *Davies* was much more numerous than *Davis* in the English counties bordering on Wales, *Davis* was far in excess in the next tier of counties.

EAST: Guppy reported *East* frequent in some of the mid-English counties. There are few cases of this surname in the 1975 study.

EDWARDS: Another surname with a Welsh pattern of distribution: 'exceedingly numerous in North and South Wales and in the adjacent English counties' (Guppy, p. 31).

For explanation of maps and graphs see p. 91

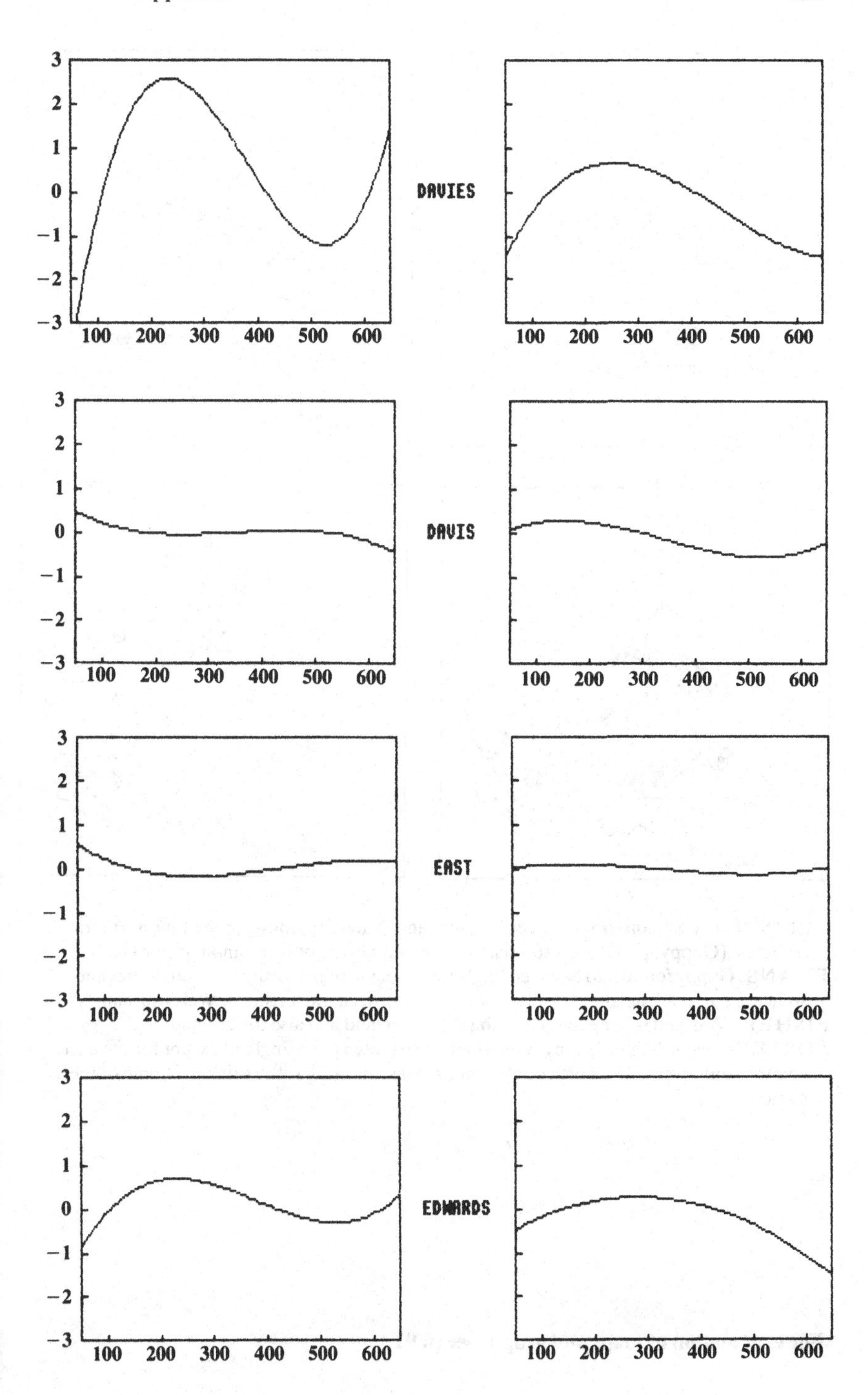
DAVIES
DAVIS
EAST
EDWARDS
3
2
1
0
−1
−2
−3
100
200
300
400
500
600

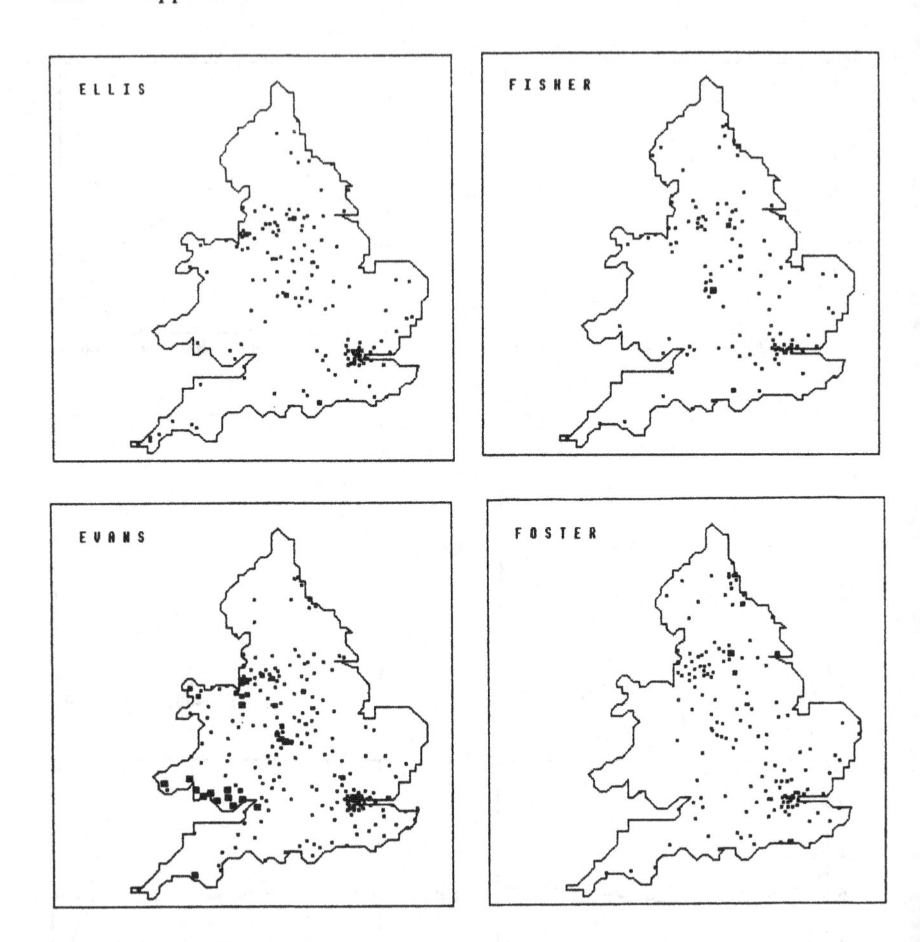

ELLIS: 'Fairly well distributed over England and Wales, excepting in the four northern counties' (Guppy, p. 32). It is still scarce in the far north but is common in Cornwall.
EVANS: Guppy found it to be exceedingly numerous in both North and South Wales and adjacent parts of England. It still shows this pattern, characteristic of Welsh surnames.
FISHER: Was and is irregularly distributed in England and also in Scotland.
FOSTER: According to Guppy, was widely distributed over England except for the east and the south-west. The north of the country was, and still is, the principal home of the name.

For explanation of maps and graphs see p. 91

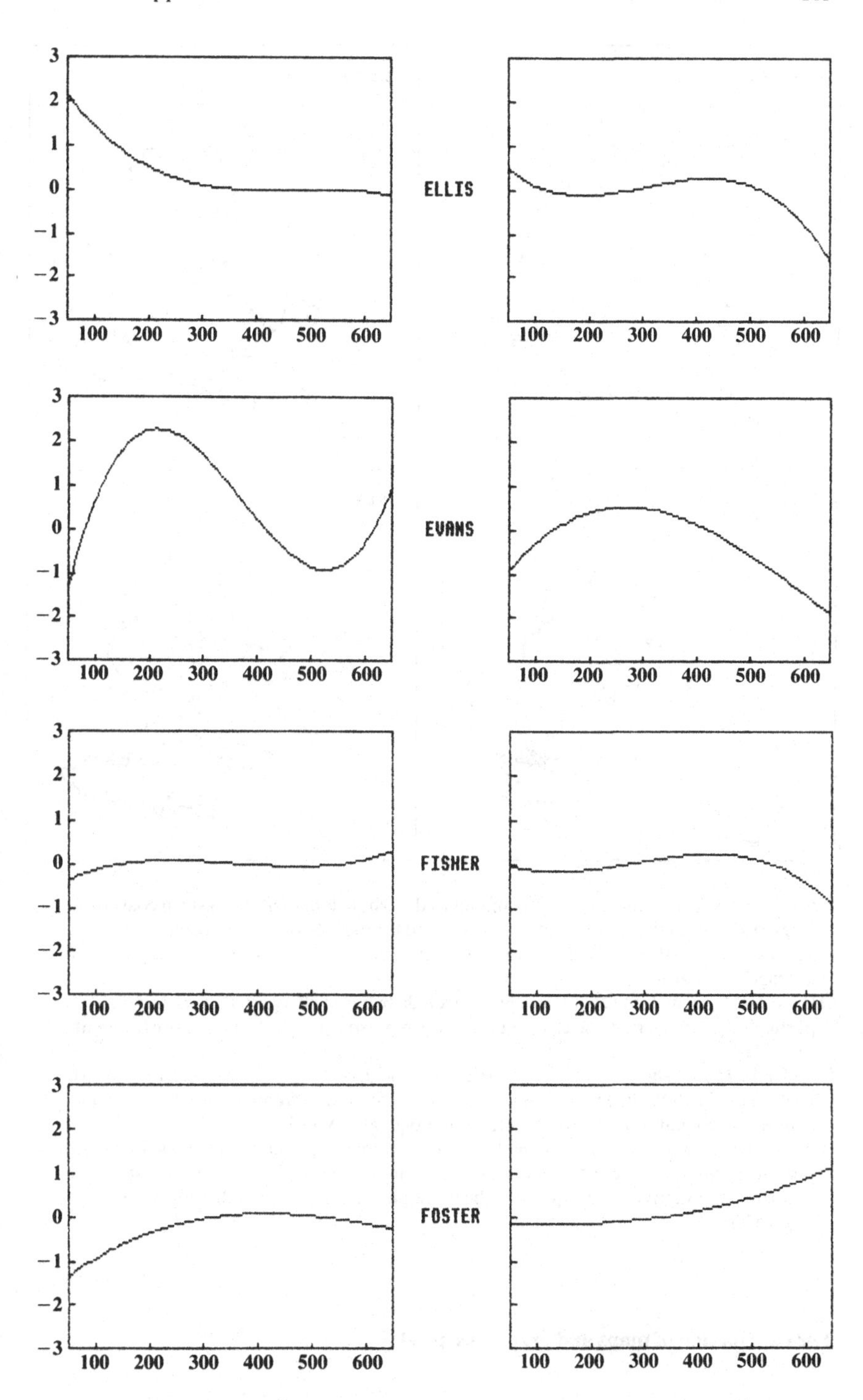
3
2
1
0
−1
−2
−3
100
200
300
400
500
600
ELLIS
EVANS
FISHER
FOSTER

GRAY: Was, generally speaking, confined to the whole length of the eastern coast of England, parts of the south coast of England and southern Scotland, according to Guppy. The extreme north-eastern and south-western corners of England still have higher than average frequencies.

GREEN: According to Guppy was pretty well distributed all over England, but particularly numerous in the east and very rare in the south-west. It is still more frequent towards the east.

GRIFFITHS: In the nineteenth century this characteristic Welsh name (often replaced by *Griffith* in North Wales) was present in about half its Welsh frequency in the adjacent counties of England. It still has a distribution typical of Welsh surnames.

HALL: Distributed over all England but with two areas of greatest frequency, one in the far north, the other in the Midlands, and rarest in the south-east and south-west, according to Guppy. Still frequent in the extreme north and rare in the south-east and south-west.

For explanation of maps and graphs see p. 91

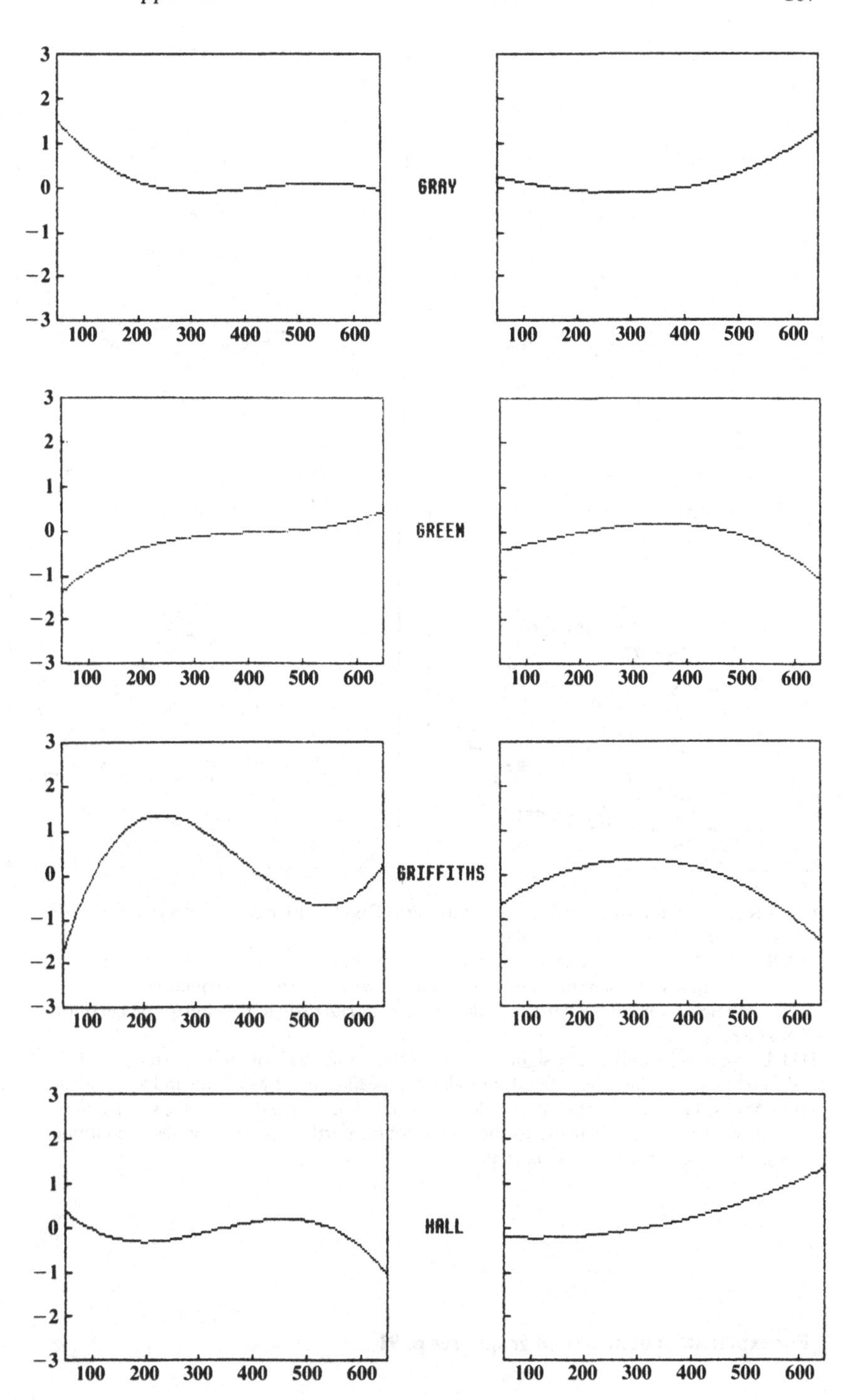

3
2
1
0
−1
−2
−3
100 200 300 400 500 600
GRAY
GREEN
GRIFFITHS
HALL

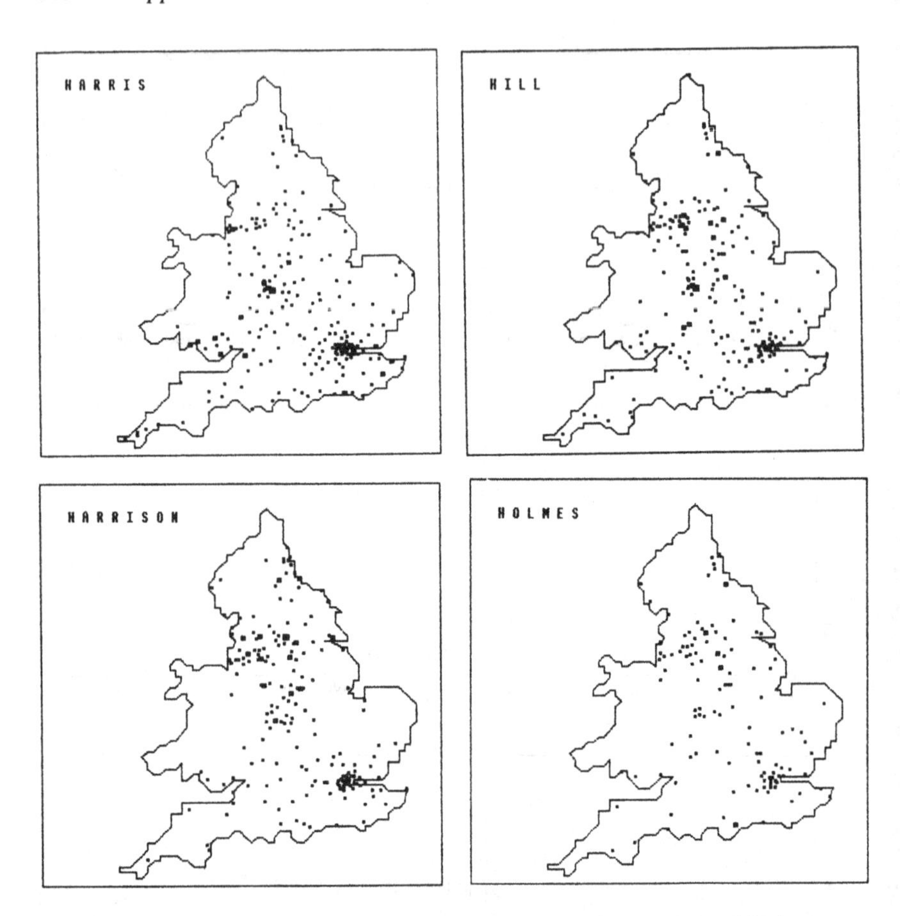

HARRIS: A southern English name according to Guppy, also frequent in South Wales. It is still sparse in the north of England.
HARRISON: It was described as especially frequent in certain areas of the north of England but also present in some of the counties where *Harris* was common. Nevertheless, we find the pattern of distribution of *Harrison* to be a mirror image of that of *Harris*.
HILL: Now lacks statistically significant clines, but at the time of Guppy's study it was described as generally distributed in England, especially in the Midlands and south-west.
HOLMES: Guppy described it as widely distributed in England except the south and south-west and common in the northern (but not the northernmost) counties. It continues to show these variations with latitude.

For explanation of maps and graphs see p. 91

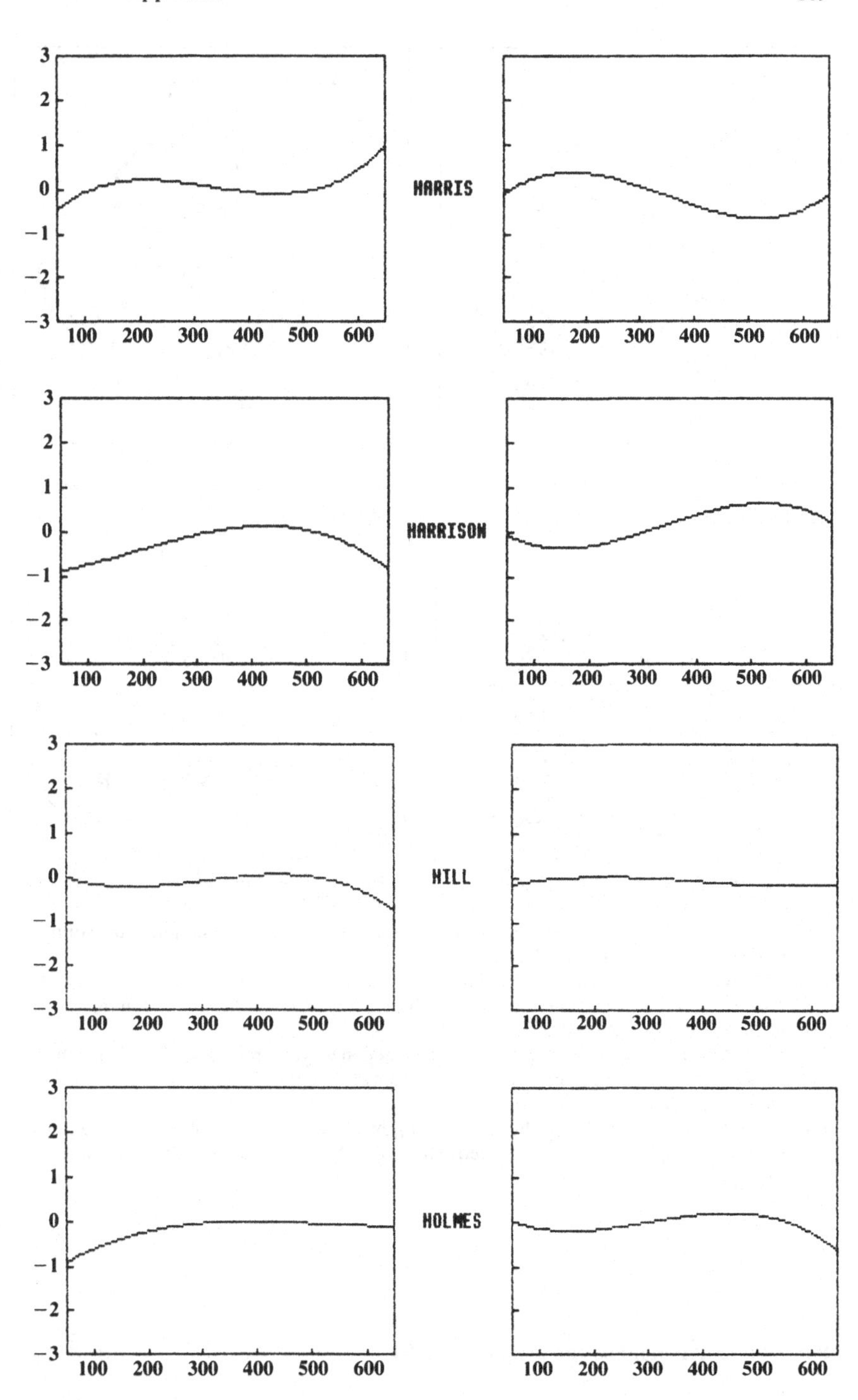
HARRIS
HARRISON
HILL
HOLMES
3
2
1
0
−1
−2
−3
100
200
300
400
500
600

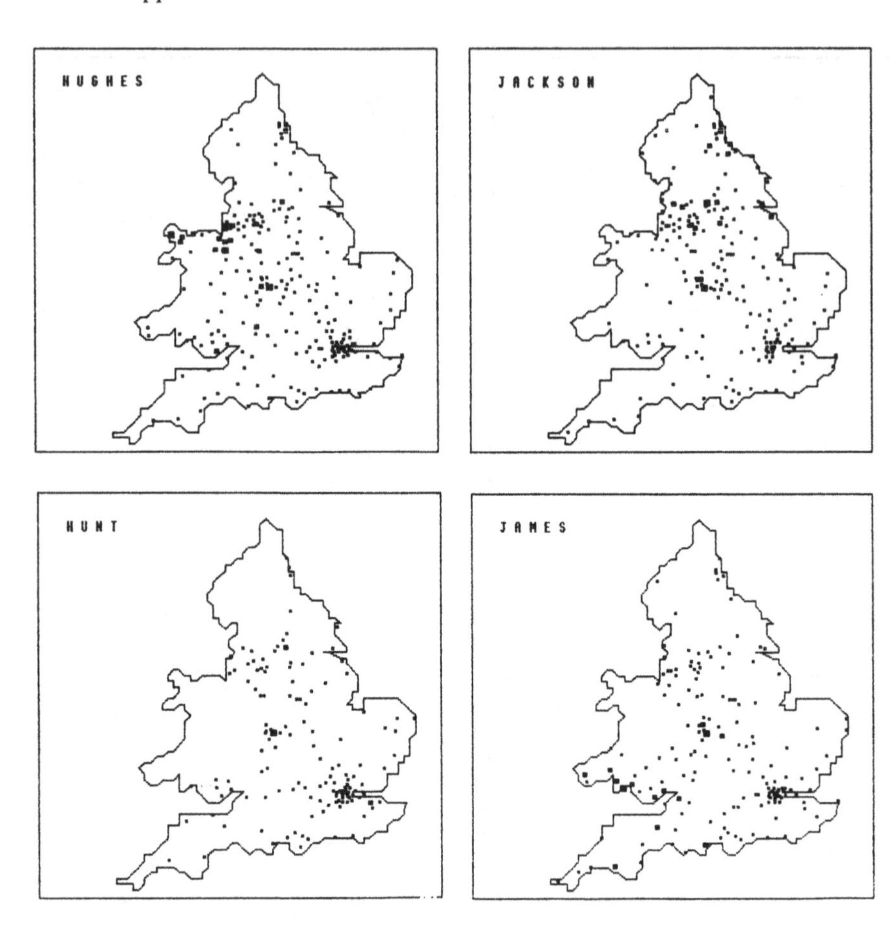

HUGHES: Was and is very frequent in North Wales but was and is common in the south of Wales and adjacent areas in England.

HUNT: 'Well distributed through England except in the north, where its place is supplied by *Hunter*, which has the same signification' (Guppy, p. 40). *Hunt* is still rare in the English counties on the border with Scotland.

JACKSON: Although found nearly all over England, it was best represented in the north and was also found in southern Scotland, according to Guppy. Today it remains predominantly northern.

JAMES: The principal home of this name in Guppy's study was South Wales and the south-west of England, but it reappeared in the north. We find it numerous in the west and south.

For explanation of maps and graphs see p. 91

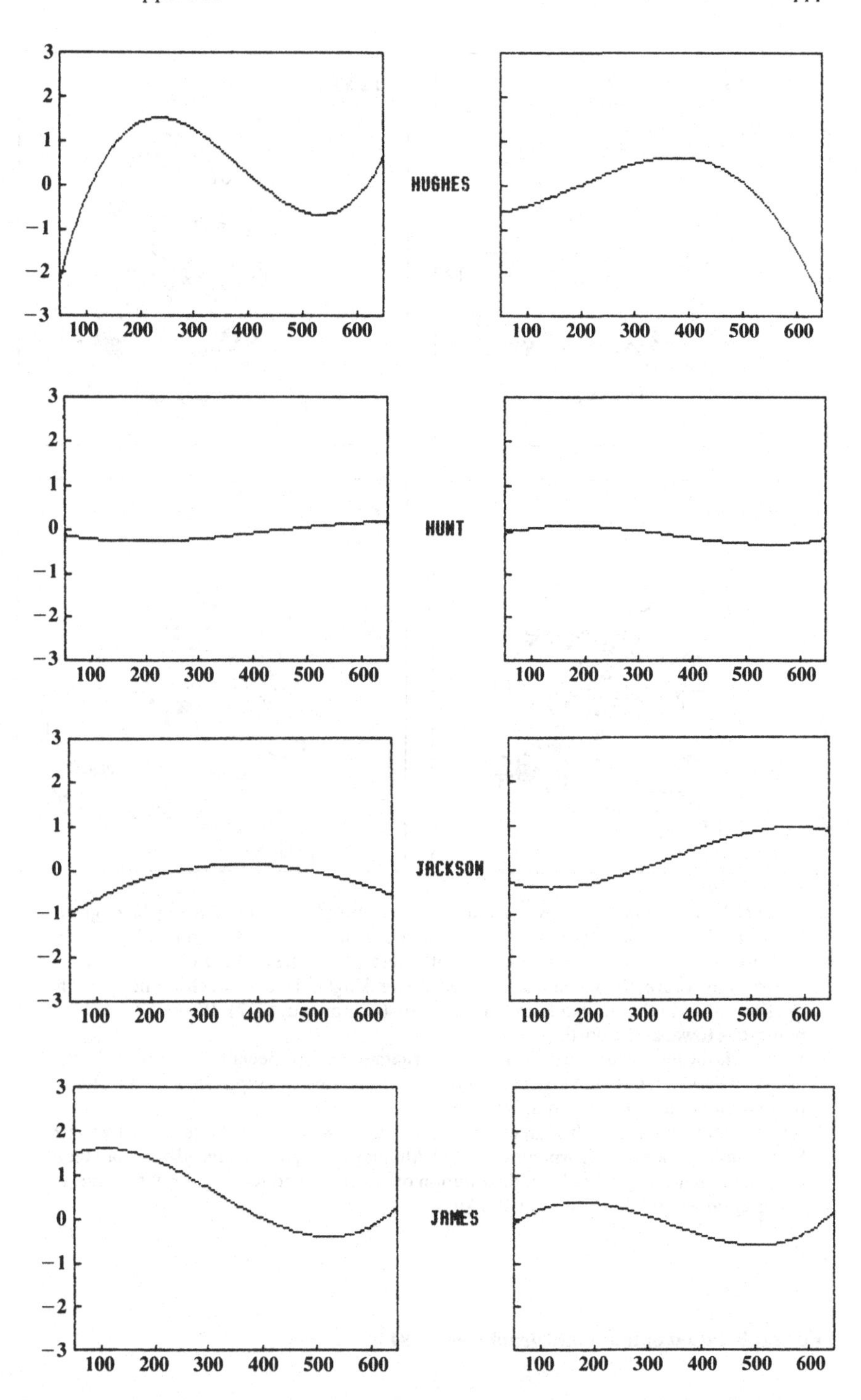
HUGHES
HUNT
JACKSON
JAMES
3
2
1
0
−1
−2
−3
100
200
300
400
500
600

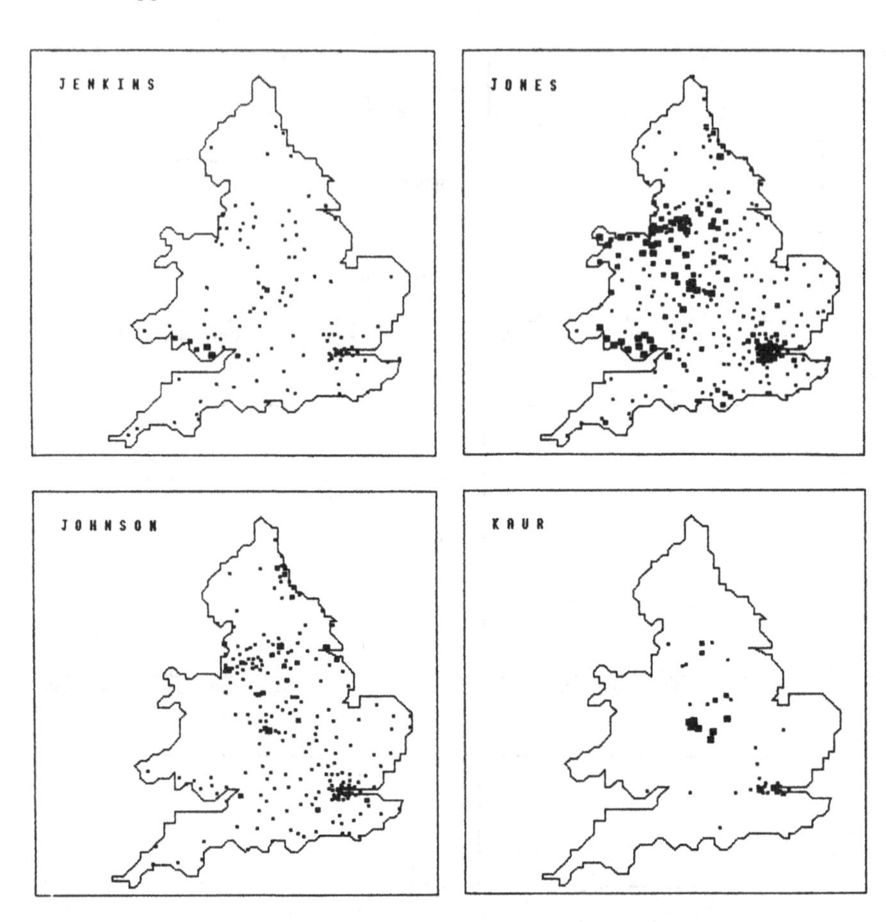

JENKINS: According to Guppy, introduced into South Wales by Flemings in the reign of Henry I. This focus and the extensions to the south and south-east remain today.

JOHNSON: 'With the exception of the south-western counties, where it is absent or conspicuously rare, this name is distributed all over England, but in much less numbers in the south than in the Midlands and in the north' (Guppy, pp. 41–2). We find it more numerous towards the north.

JONES: 'Is the most characteristic of Welsh surnames, being especially frequent in North Wales, where one out of every seven persons is thus named. (Guppy, p. 42). It is still prominent in the same and adjacent areas.

KAUR: Not listed in British sources, this Sikh name is used in lieu of a surname by some Sikh women. Since use of surnames among Sikhs may in part be influenced by social status or date of migration to England, distribution of women called *Kaur* may not be entirely representative of the distribution of Sikhs.

For explanation of maps and graphs see p. 91

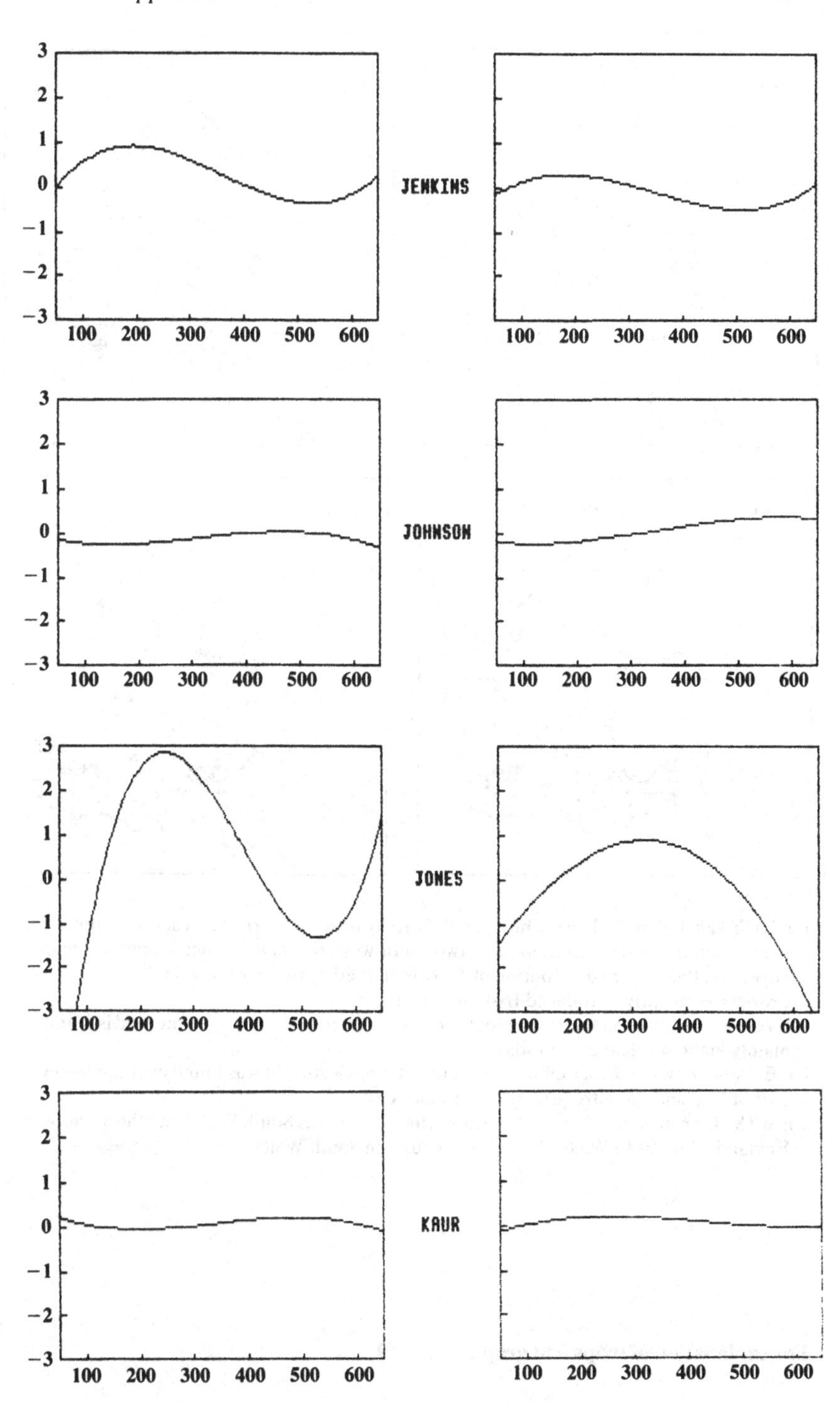
JENKINS
JOHNSON
JONES
KAUR
3
2
1
0
−1
−2
−3
100
200
300
400
500
600

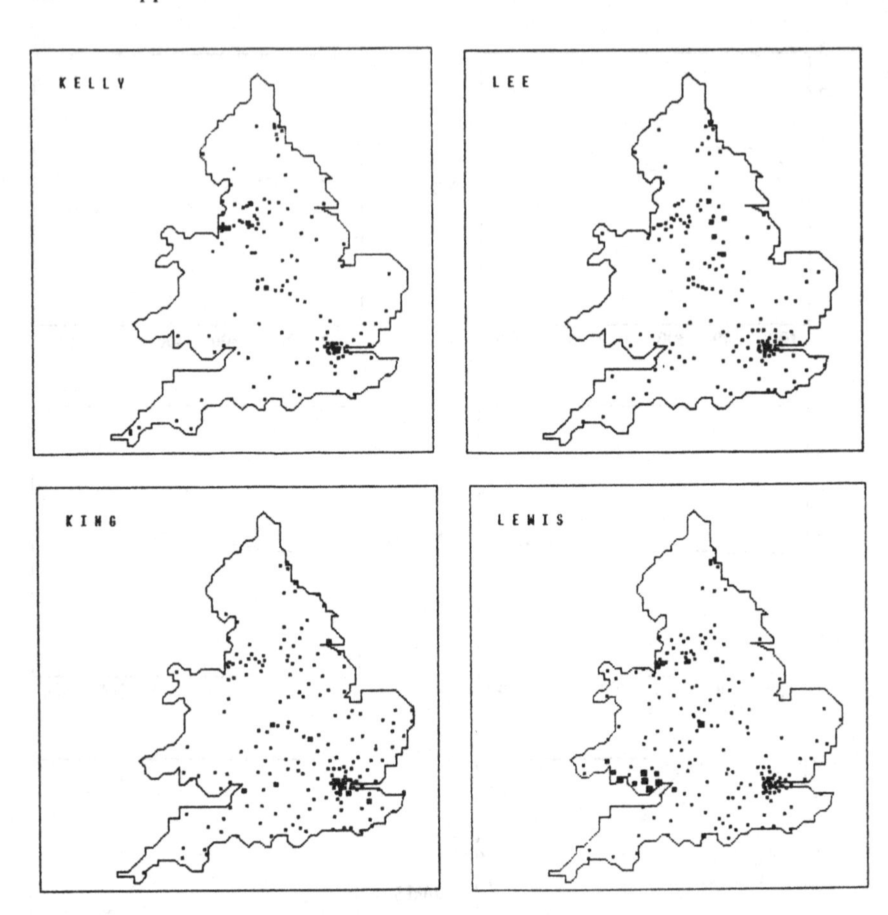

KELLY: Seated in the Devonshire parish of Kelly near the Cornish border since the twelfth century and frequent in these two south-western counties when Guppy wrote. However, the present distribution of *Kelly* is marked by its association in the same districts as recently introduced Irish surnames.
KING: Was mostly confined to the southern half of England in Guppy's time, and is found mainly in the south and east today.
LEE: Now shows no geographic clines, but in Guppy's study it was found over the larger part of England but infrequently in the south-east.
LEWIS: Is an ancient Welsh name whose chief centre was South Wales and the areas of England adjacent to Wales. Its present focus is in South Wales.

For explanation of maps and graphs see p. 91

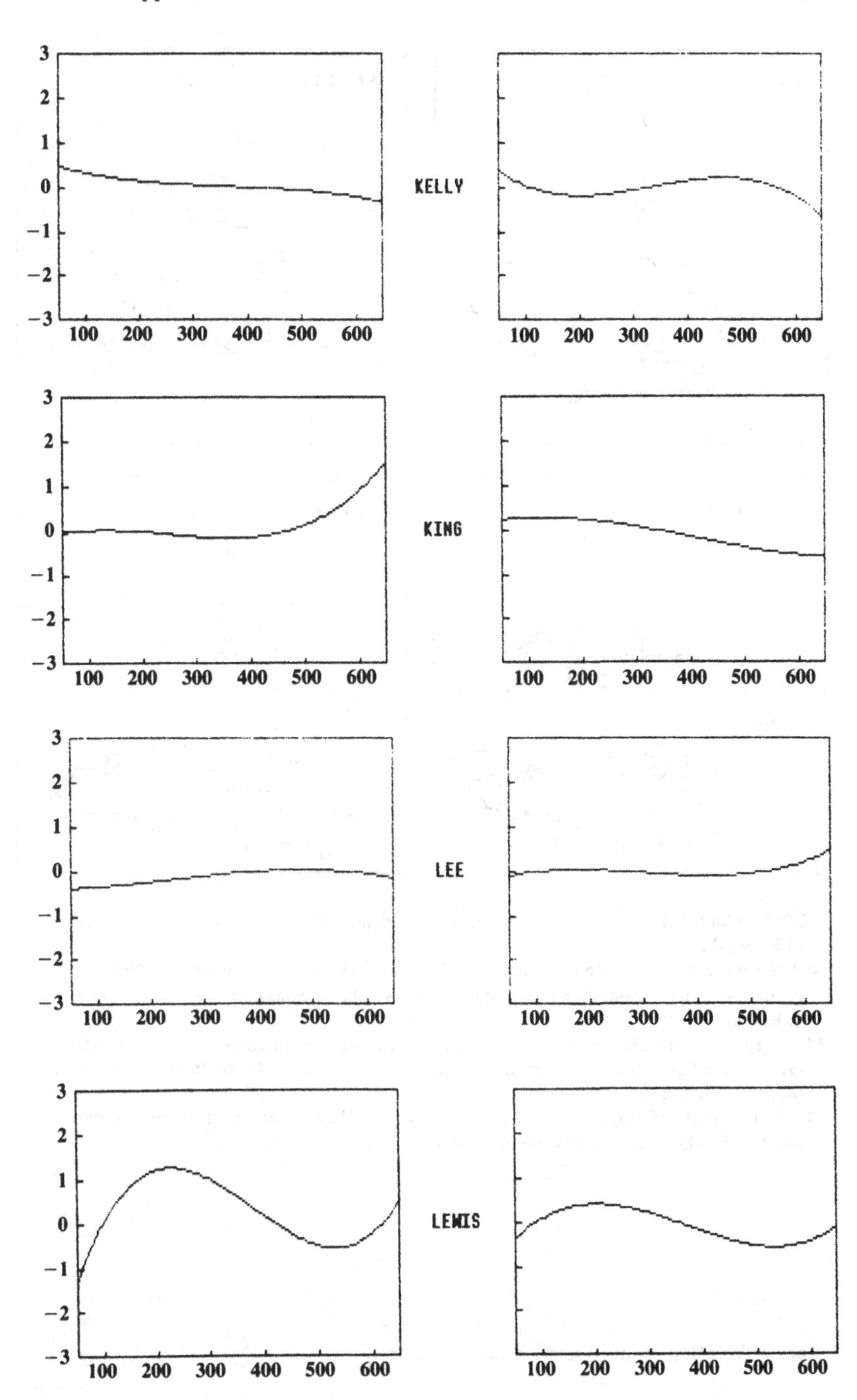
3
2
1
0
−1
−2
−3
100 200 300 400 500 600
KELLY
KING
LEE
LEWIS

LONG: Confined largely to the south in both 1890 and 1975; it was numerous in the reign of Edward I.

MARSHALL: The name is usually considered to derive from 'mare-schalks' (horse groom or farrier) rather than from the French 'marechal'. It was and is distributed widely in England.

MARTIN: Distributed over the whole of England and southern Scotland according to Guppy. We find a general linear decline in frequency towards the north and more cases than expected in the south-west.

MASON: Said by Guppy to be scarcely represented in the extreme south and the extreme north of England, but we find it common in the north.

For explanation of maps and graphs see p. 91

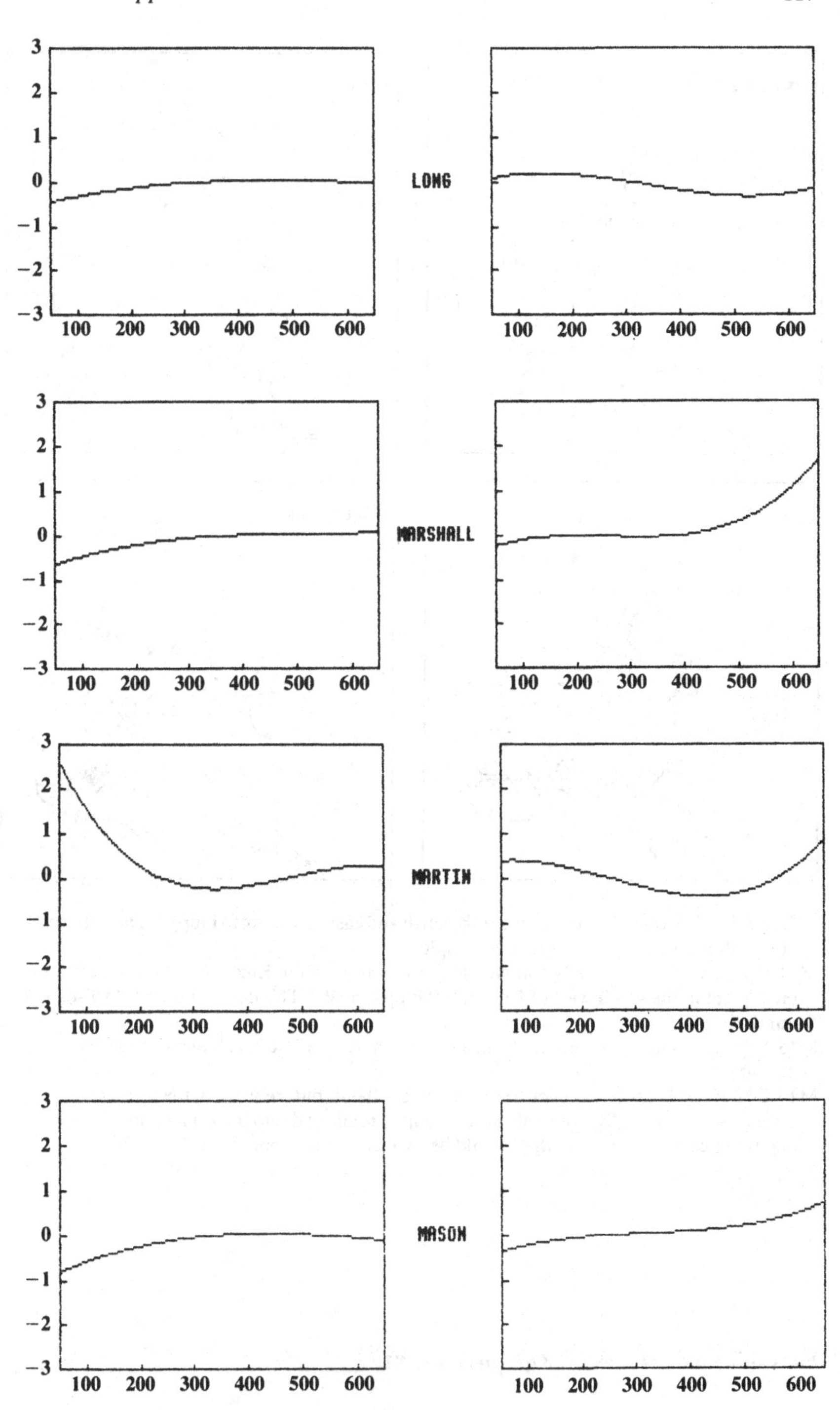

LONG
MARSHALL
MARTIN
MASON
3
2
1
0
−1
−2
−3
100
200
300
400
500
600

MILLER: Centres occurred in the south, north and east according to Guppy. The pattern of distribution does not appear to be simple.

MILLS: 'This name is mostly confined to the southern half of England . . . It is rare or infrequent in the south-west of England' (Guppy, p. 46). This description would also fit our findings.

MISTRY: This name occurs chiefly in a few northern and Midland industrial districts and in London.

MITCHELL: Distributed over England and Scotland, but frequent in the south of England and especially Cornwall, where Guppy remarked it to be prominent. We also find more cases in Cornwall than would be expected at random.

For explanation of maps and graphs see p. 91

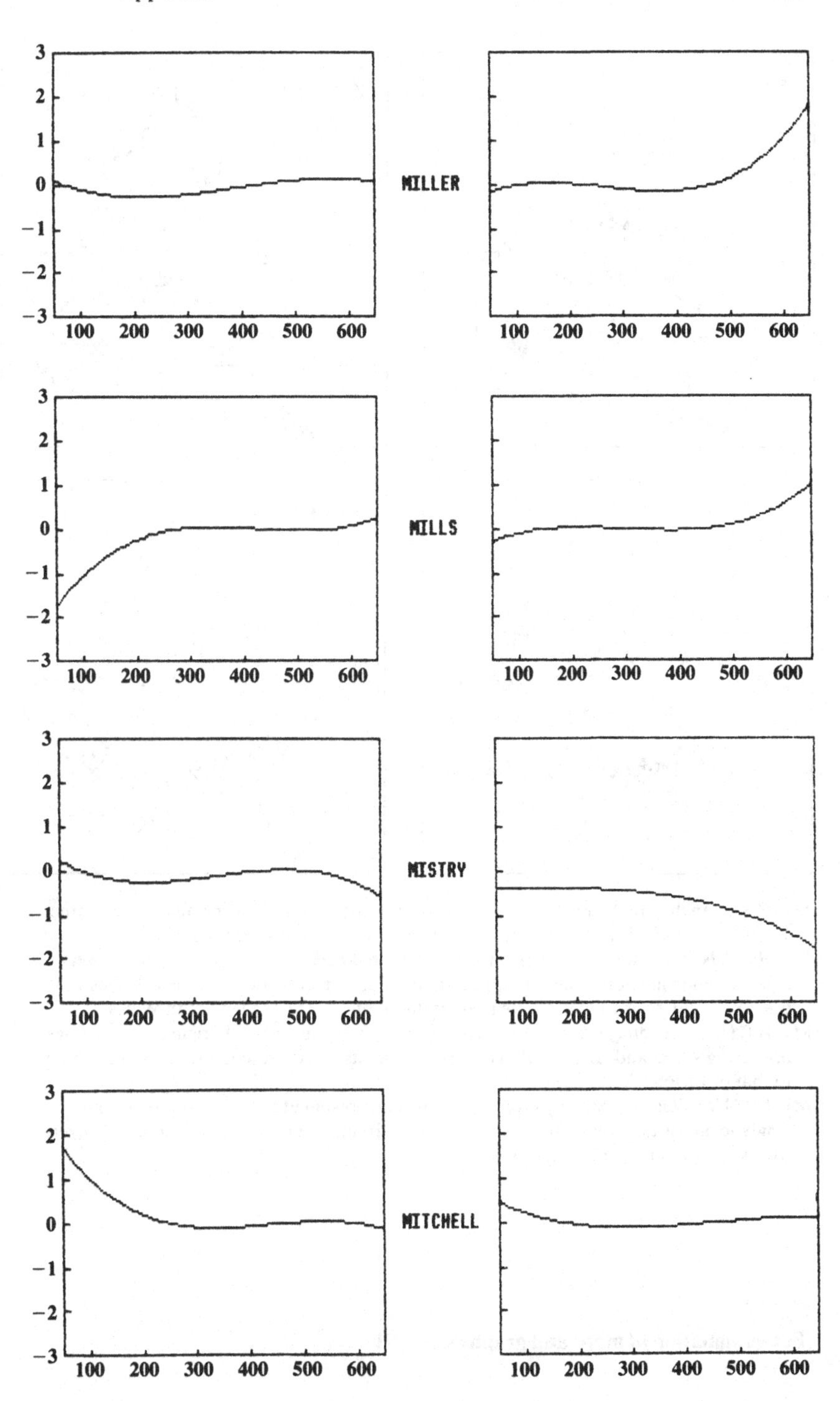
MILLER
MILLS
MISTRY
MITCHELL
3
2
1
0
−1
−2
−3
100 200 300 400 500 600

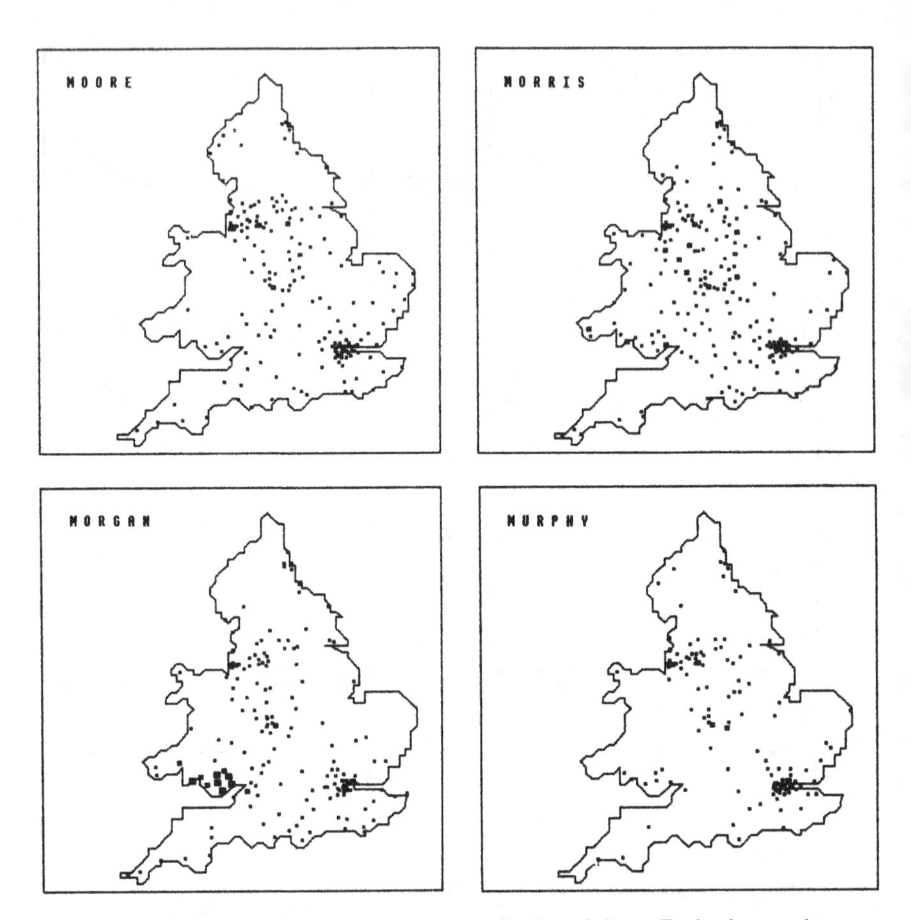

MOORE: In the nineteenth century this name was distributed all over England except the south coast where it was uncommon. There are now no statistically significant clines.
MORGAN: Is an ancient Welsh personal name that became a surname about 500 years ago. Common in South Wales, Guppy recorded 230 farmers called *Morgan* in Wales and the border shires of England. The present distribution is centred in South Wales.
MORRIS: According to Guppy, *Morris* had more than one centre of origin in the counties bordering Wales and, afterwards, centres in Wales itself. Its present distribution is similar to that of various Welsh surnames.
MURPHY: Was not listed by Guppy and must be presumed to have become common in England only after large-scale immigration from Ireland since the potato famine. *Murphy* shows no clines in England and Wales.

For explanation of maps and graphs see p. 91

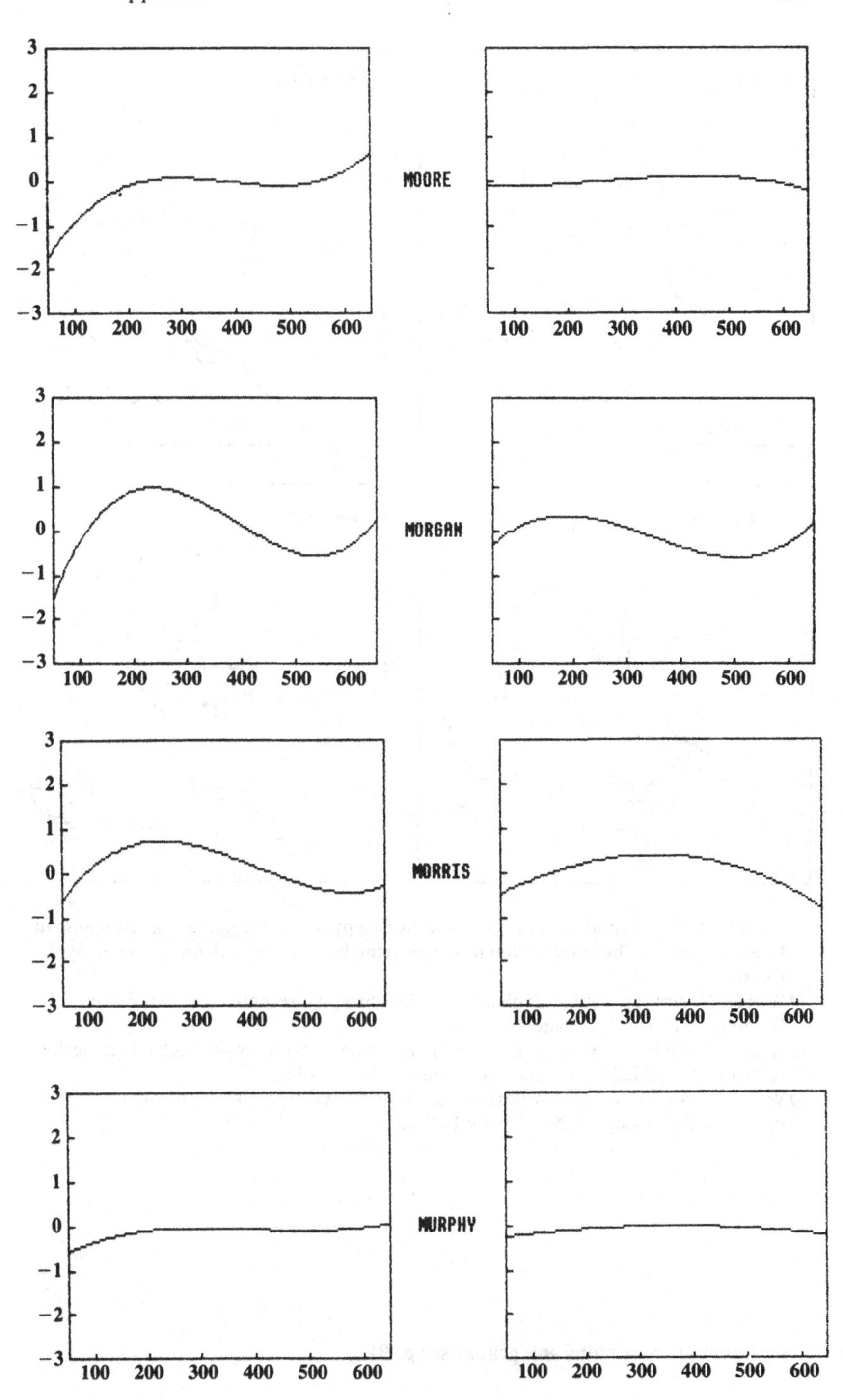
3
2
1
0
−1
−2
−3
100 200 300 400 500 600
MOORE
MORGAN
MORRIS
MURPHY

NORTH: Chiefly reported in the Midlands by Guppy, it was the name of a distinguished Leicester family. There are too few instances in the 1975 sample to define any geographic trend.
O'BRIEN: Not reported in England in earlier times, it is recently introduced from Ireland. No clines are apparent.
O'NEILL: An Irish name, recent in England, but *Neel* was present in England during the reign of Edward I. *O'Neill* now has a generally flat distribution.
OWEN: Mostly confined to Wales, especially North Wales, according to Guppy. This pattern is also strongly defined in the 1975 data.

For explanation of maps and graphs see p. 91

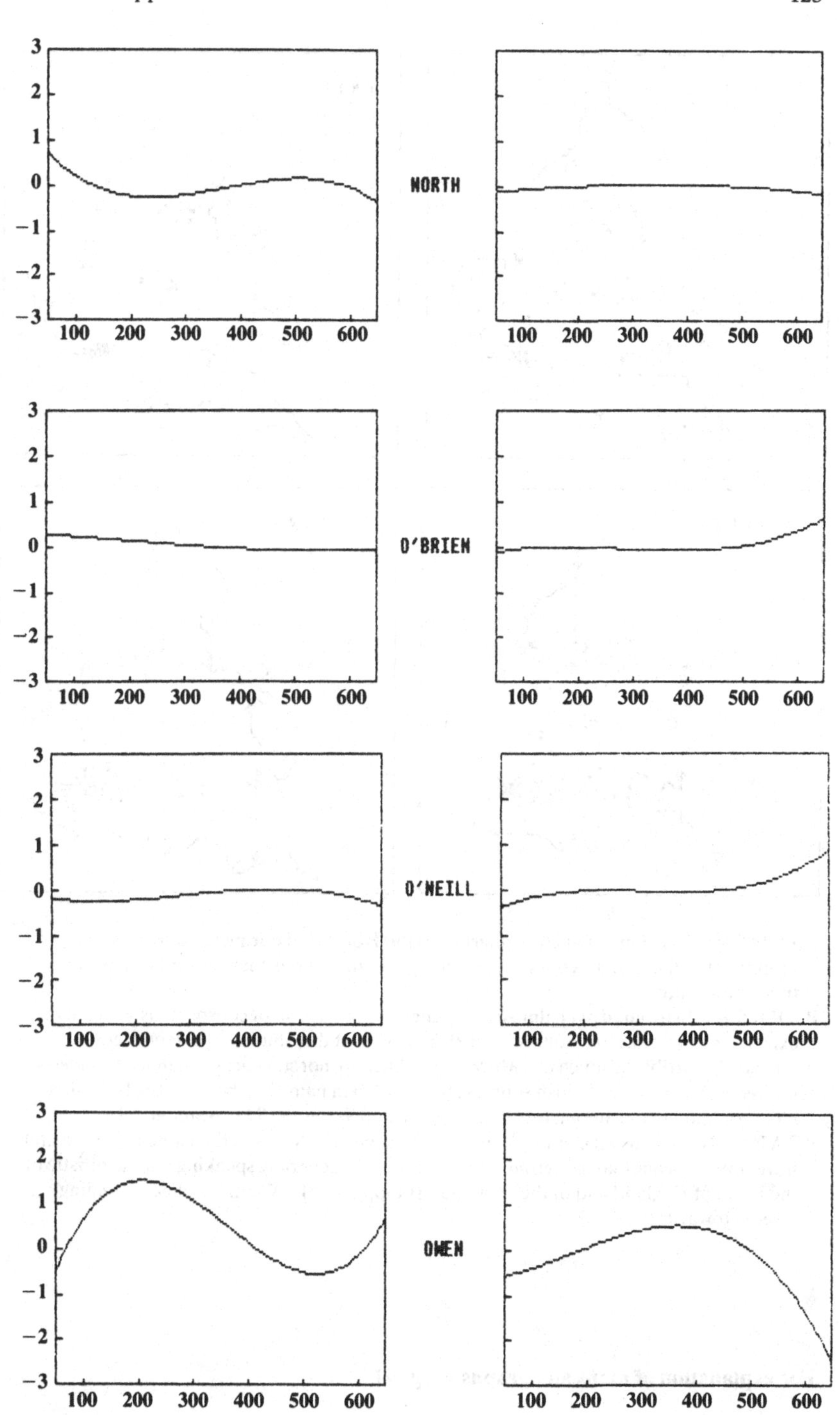
NORTH
O'BRIEN
O'NEILL
OWEN
3
2
1
0
−1
−2
−3
100
200
300
400
500
600

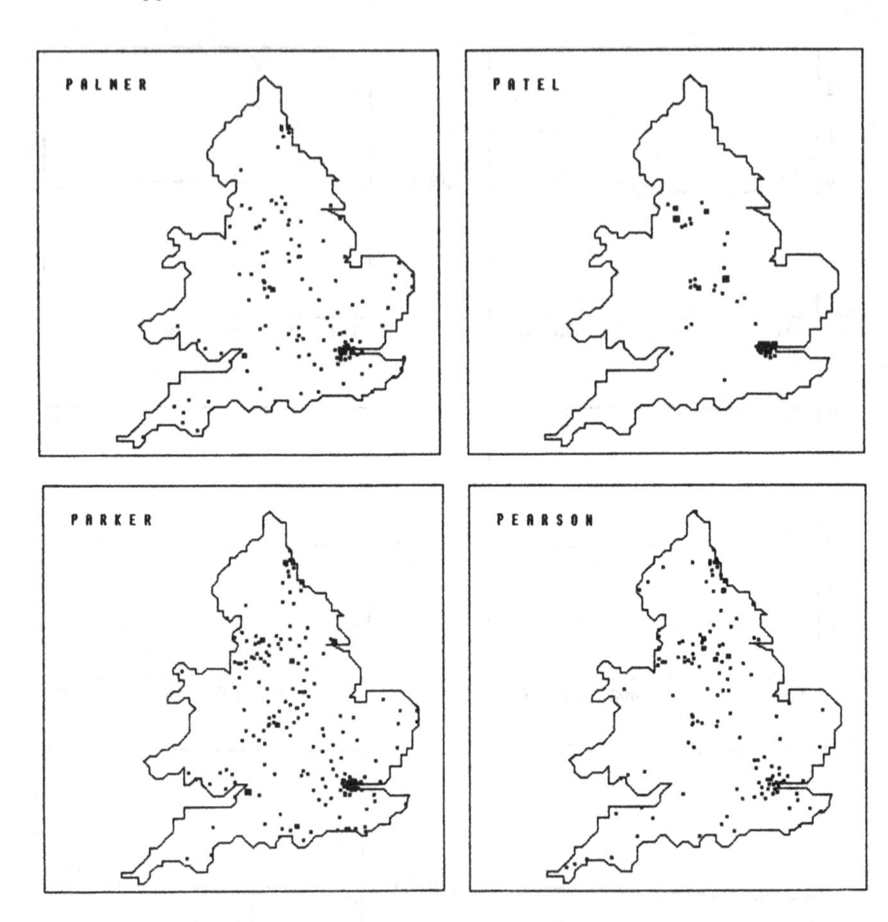

PALMER: This name of ancient pilgrims to the Holy Land during the crusades is prominent in the eastern counties, as Guppy noted, and was there as far back as the thirteenth century.

PARKER: Distributed over almost all of England, but absent or conspicuously rare in the extreme south-western counties in 1890. The present distribution is homogeneous.

PATEL: Is distributed much like *Mistry*. The clines are not statistically significant. *Patel* is not recorded in earlier British sources, but this Asian name had become the forty-first most common name among births in England and Wales in 1975 (Ammon, 1976).

PEARSON: 'The usual distinction prevails between the forms of the name that have and have not the Scandinavian termination of "son" . . . generally speaking, characteristic of the north of England and of the Midlands' (Guppy, p. 51). We find a cline ascending steeply towards the north.

For explanation of maps and graphs see p. 91

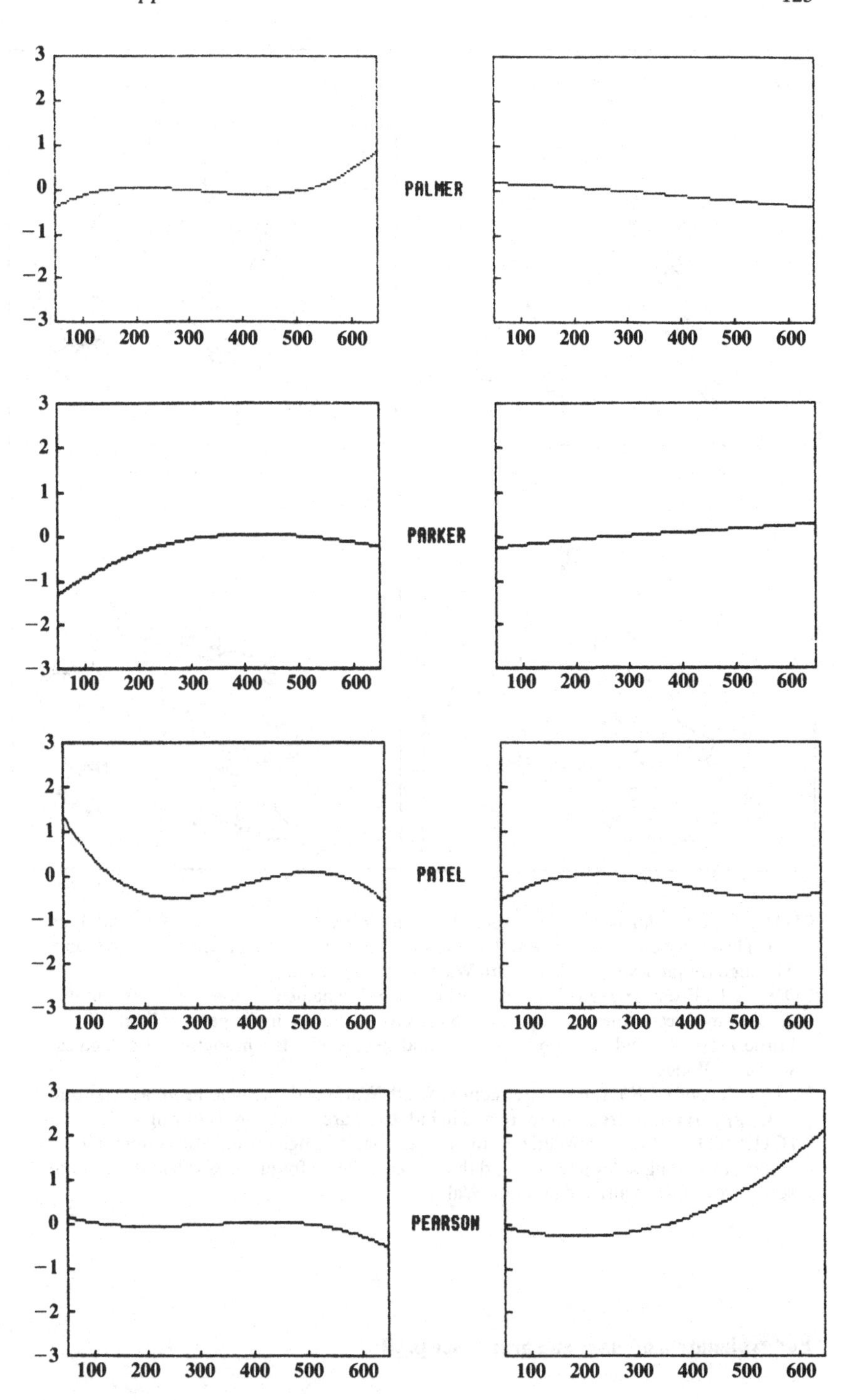
PALMER
PARKER
PATEL
PEARSON
3
2
1
0
−1
−2
−3
100
200
300
400
500
600

PHILLIPS: Confined to Wales, especially South Wales, and the south of England, being much less frequent in the eastern than the western part of this area, according to Guppy. The high frequency in south-western Wales is still apparent.

POWELL: From *Ap Howel*, it had only been a Welsh name for three centuries, but its great home, according to Guppy, was in Herefordshire and in the parts of England bordering Wales and in South Wales. We find instances widely in southern England as well as in Wales.

PRICE: From *Ap Rhys*, it was frequent in South Wales and surrounding areas according to Guppy. It is now frequent there and in industrial areas of the west of England.

RICHARDS: Mostly crowded into the western half of England and also common in Wales, according to Guppy. We find the highest relative frequencies still in the extreme south-west (Cornwall) and in South Wales.

For explanation of maps and graphs see p. 91

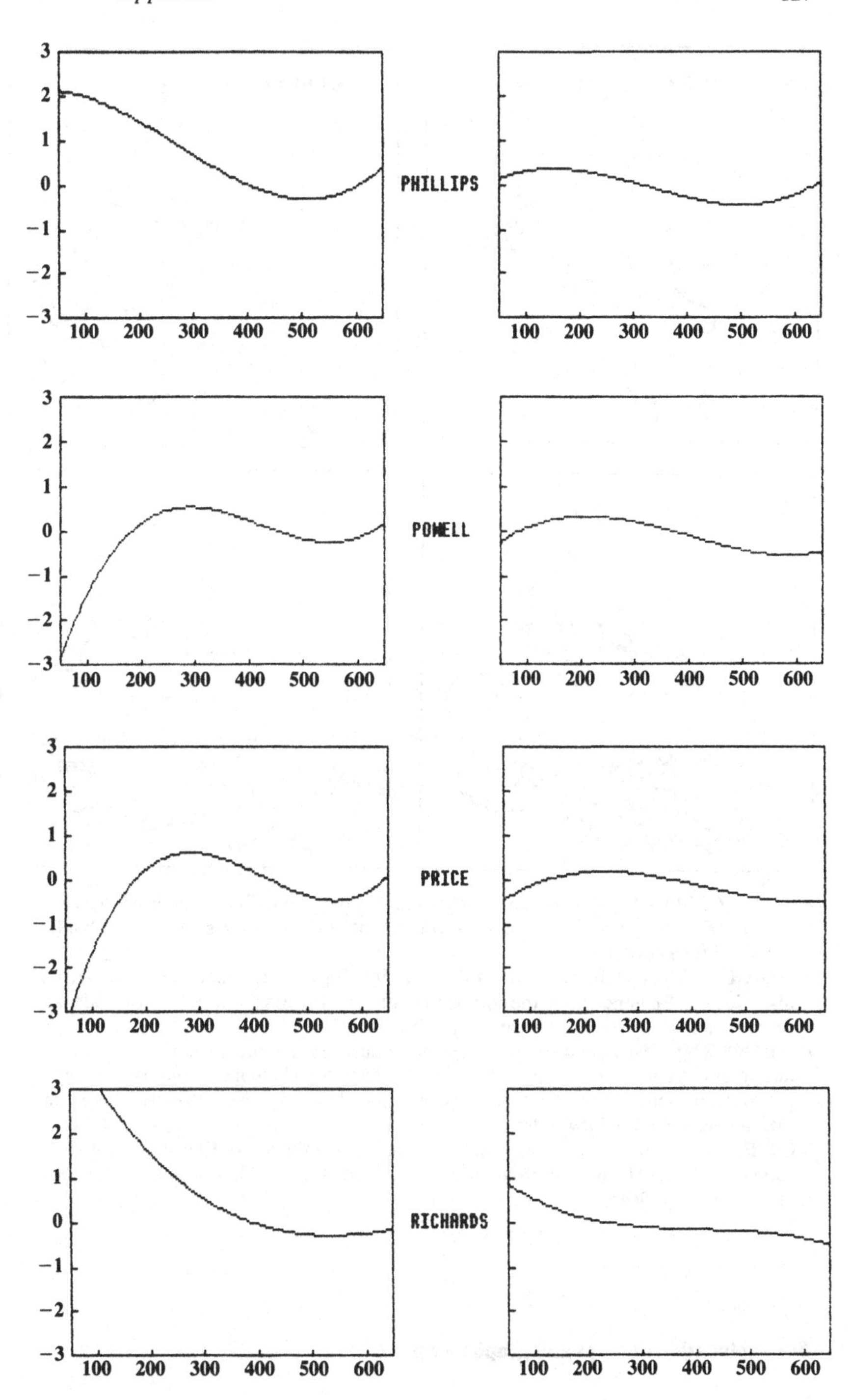
PHILLIPS
POWELL
PRICE
RICHARDS
3
2
1
0
−1
−2
−3
100 200 300 400 500 600

RICHARDSON: Now definitely more frequent in the north, Guppy reported that it replaced *Richards* in north and east coast counties of England and was found on both sides of the Scottish border.

ROBERTS: Mostly found in North Wales and adjoining regions, according to Guppy, but also widely in England except the northern counties and rare in the east. We also find that it predominates in North Wales and adjoining areas.

ROBINSON: 'Distributed over England, except in the south-west, where it is either absent or extremely rare. Its great home is in the northern half of the country' (Guppy, p. 56). Its present strongly northern concentration also means that it is most frequent at intermediate English longitudes.

ROGERS: Rare in northern England but scattered over the rest of England and Wales, though infrequent in the eastern counties, according to Guppy. The present clines are not statistically significant.

For explanation of maps and graphs see p. 91

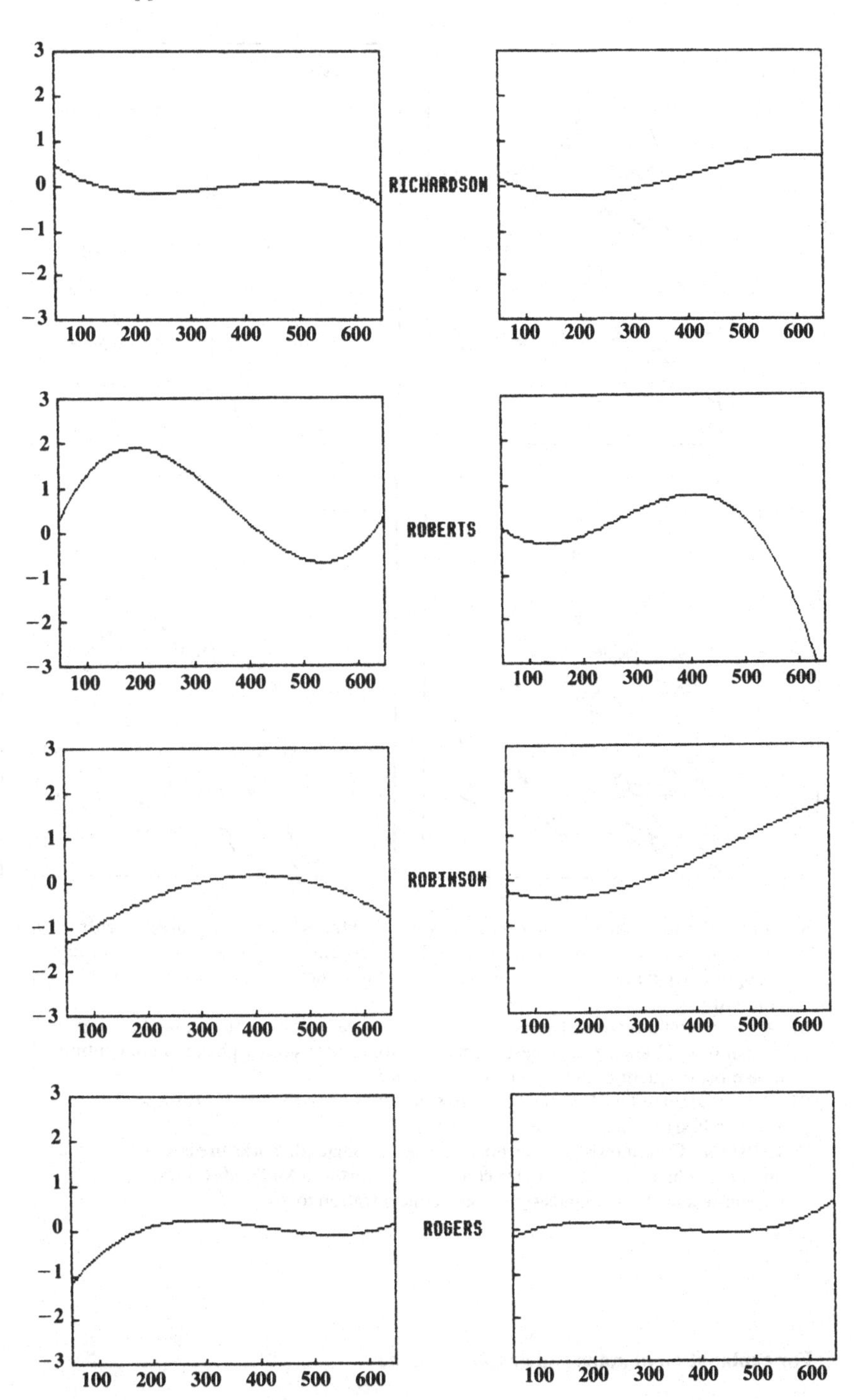
RICHARDSON
ROBERTS
ROBINSON
ROGERS
3
2
1
0
−1
−2
−3
100
200
300
400
500
600

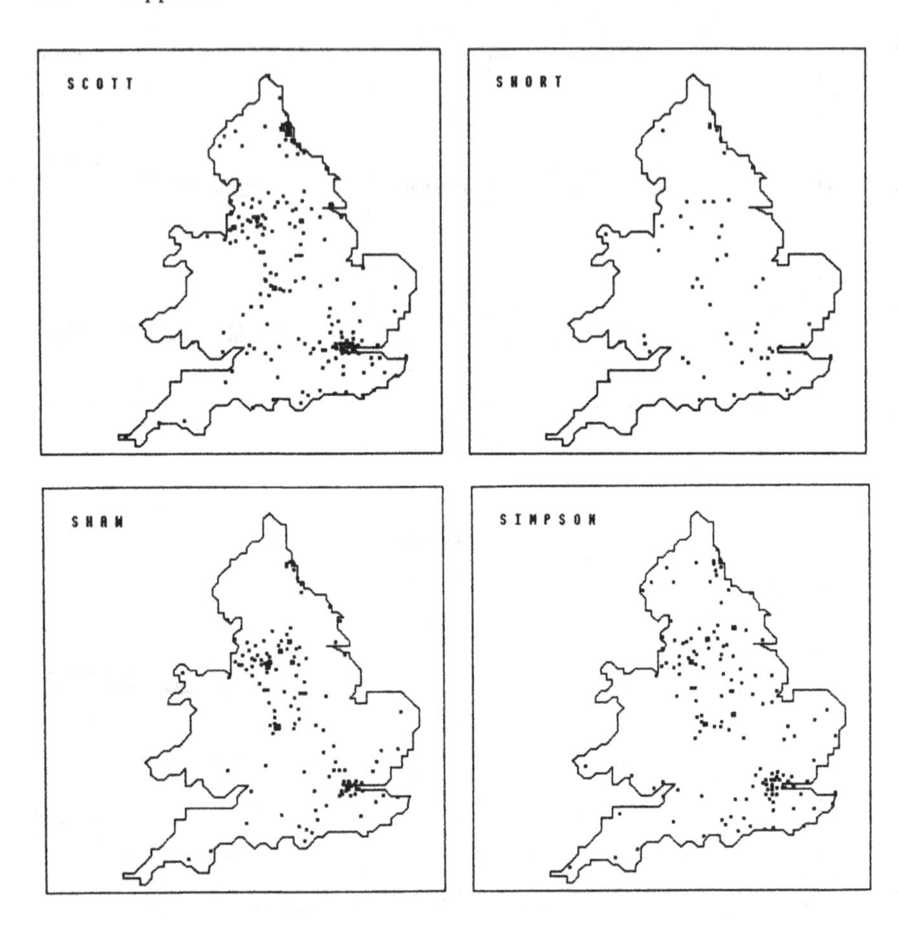

SCOTT: Shows a sharply higher frequency in the extreme north. Its principal centre was in the counties on both sides of the Scottish border, but as it also occurred scattered throughout England and in the south it may not have derived from the name of the nationality (Guppy).

SHAW: In Anglo-Saxon signified a small wood; it was common, according to Guppy, in Yorkshire and Lancashire, where the name is attached to several places. *Shaw* continues to be most frequent at the latitude of Yorkshire.

SHORT: Occurred in Bideford, Devon, since the sixteenth century. It shows no statistically significant clines in frequency now.

SIMPSON: 'Characteristic of the northern half of England. Yorkshire is its great home . . . eminently a name of the northern counties and northern Midlands (Guppy, p. 59). Yorkshire and the Midlands show some concentration today.

For explanation of maps and graphs see p. 91

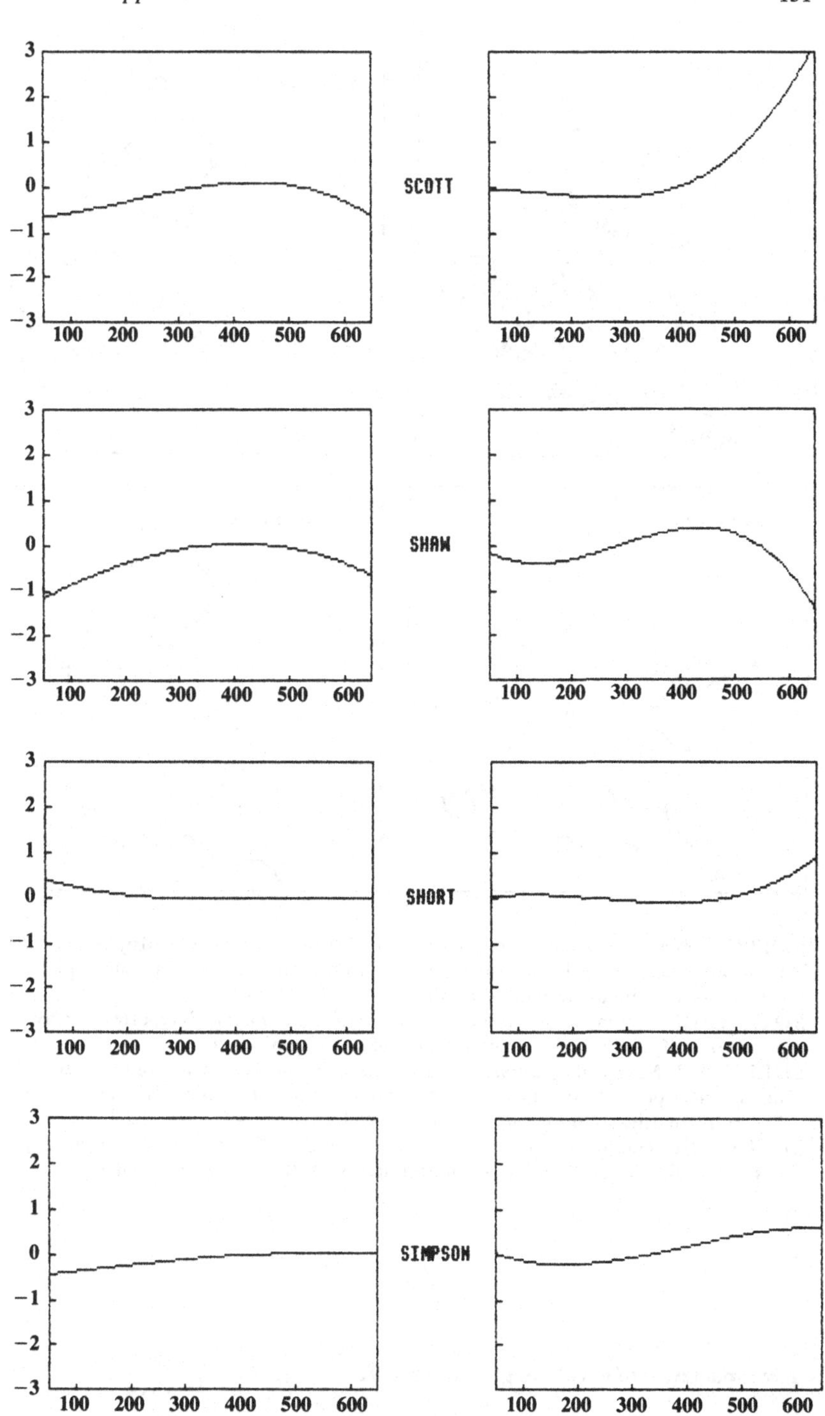
SCOTT
SHAW
SHORT
SIMPSON
3
2
1
0
−1
−2
−3
100
200
300
400
500
600

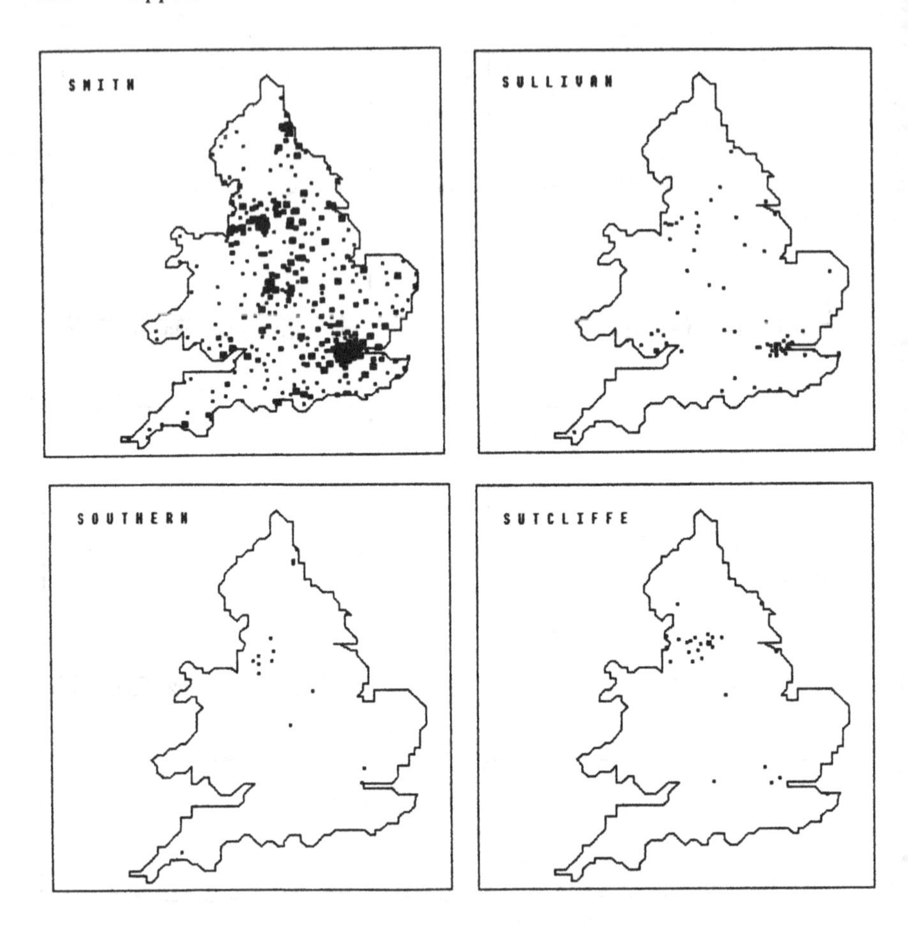

SMITH: Now most frequent in the east and at middle English latitudes. Said by Guppy to be generally distributed but infrequent in the extreme south-west and in Wales; also rather less frequent in the extreme north.

SOUTHERN: Not mentioned by Guppy, but *South*, a rarer name, is reported. Today *Southern* is, if anything, more northern in England than southern.

SULLIVAN: Not listed by Guppy, and instances in England and Wales must be recent immigrants from Ireland. Unlike the other Irish surnames, there is a significant cline of higher frequencies southward in England and Wales.

SUTCLIFFE: Guppy lists it as common in the West Riding of Yorkshire, and specifically in Halifax; also in Lancashire. It is concentrated in just these areas at present.

For explanation of maps and graphs see p. 91

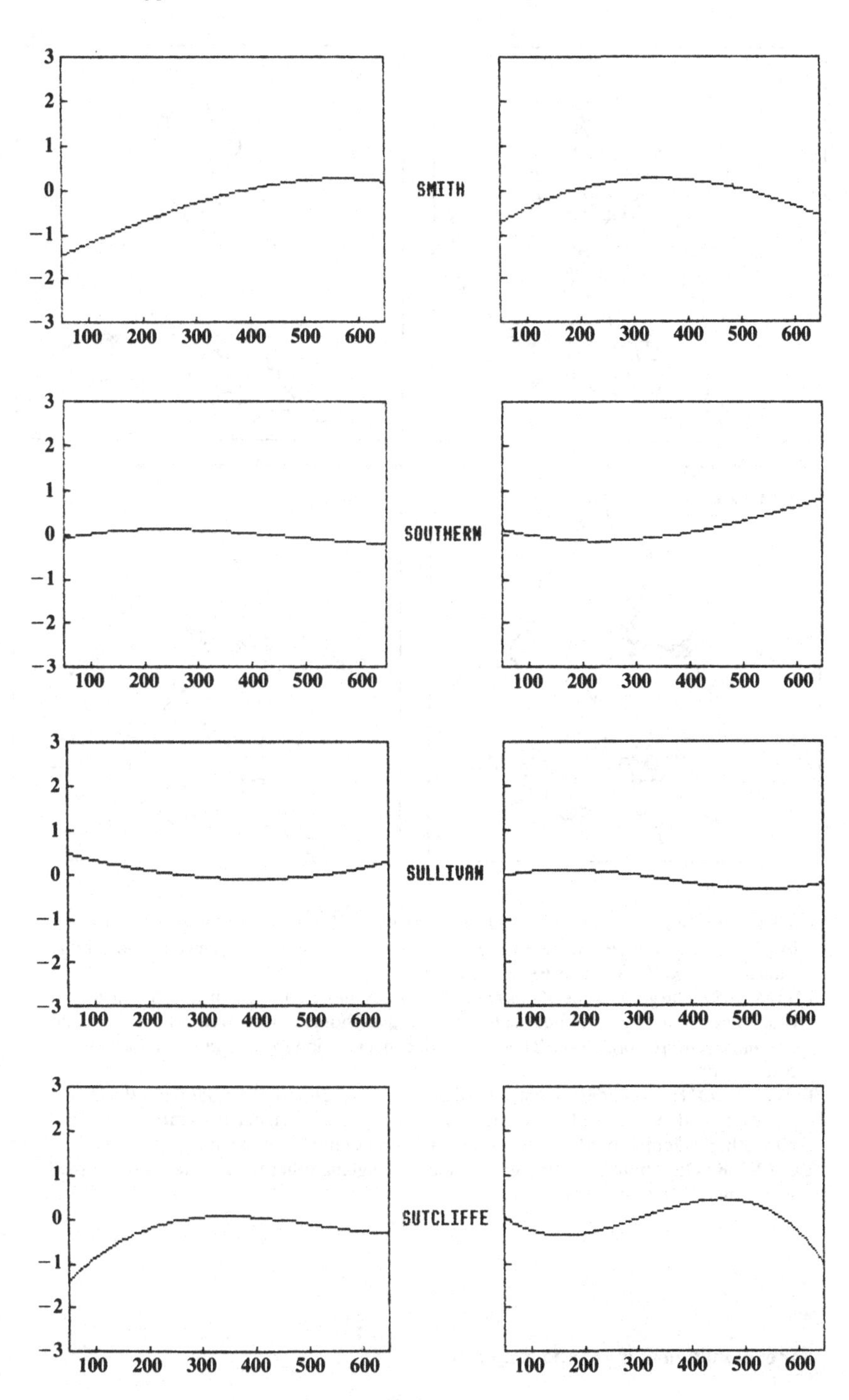
SMITH
SOUTHERN
SULLIVAN
SUTCLIFFE
3
2
1
0
−1
−2
−3
100
200
300
400
500
600

TAYLOR: Guppy wrote that it was distributed all over England, but comparatively scarce in south coast counties and near the Scottish border and in Wales. Concentrations in Lancashire and Yorkshire are still apparent.
THOMAS: 'The great home of this name is in Wales, more especially in South Wales . . . It has, however, a secondary but evidently an original home in Cornwall, where the name abounds' (Guppy, pp. 60–1). These areas of concentration show in our map of the distribution.
THOMPSON: 'This name is distributed over the greater part of England . . . It is, however, well represented in the Midlands. Further south . . . it becomes rare or dies out altogether' (Guppy, p. 61). This surname still predominates in the north.
TURNER: Distributed over the greater part of England other than the far north in 1890; very generally distributed now.

For explanation of maps and graphs see p. 91

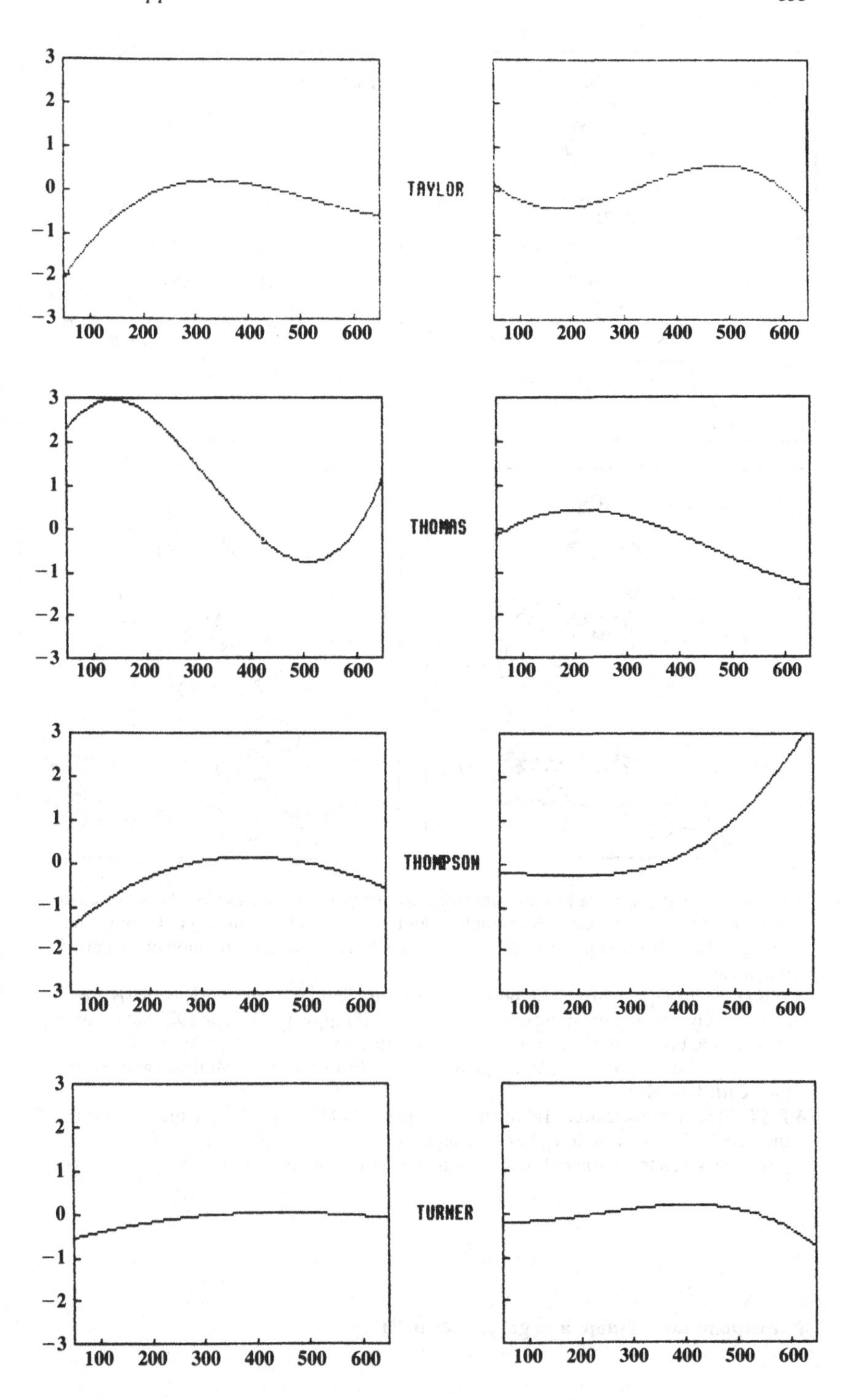

3
2
1
0
−1
−2
−3
100 200 300 400 500 600
TAYLOR
THOMAS
THOMPSON
TURNER

WALKER: This surname has the same significance as *Fuller* and *Tucker*. *Walker* was the form most general in the north of England and the Midlands, according to Guppy, and also general in Scotland, except the extreme north. It is still most common in England in the north.

WARD: 'Is infrequent in the four northernmost counties of England and is similarly absent or relatively uncommon in the southern counties' (Guppy, p. 62). The 1975 data show a tendency to higher frequencies towards the north-east.

WATSON: This name was and is frequent in the north and northern Midlands and across the Scottish border.

WEST: 'This name is scattered about in different parts of England, both in the west and in the east, and its distribution gives only slight support to the suggestion that it was originally given to persons who came from the west' (Guppy, p. 64).

For explanation of maps and graphs see p. 91

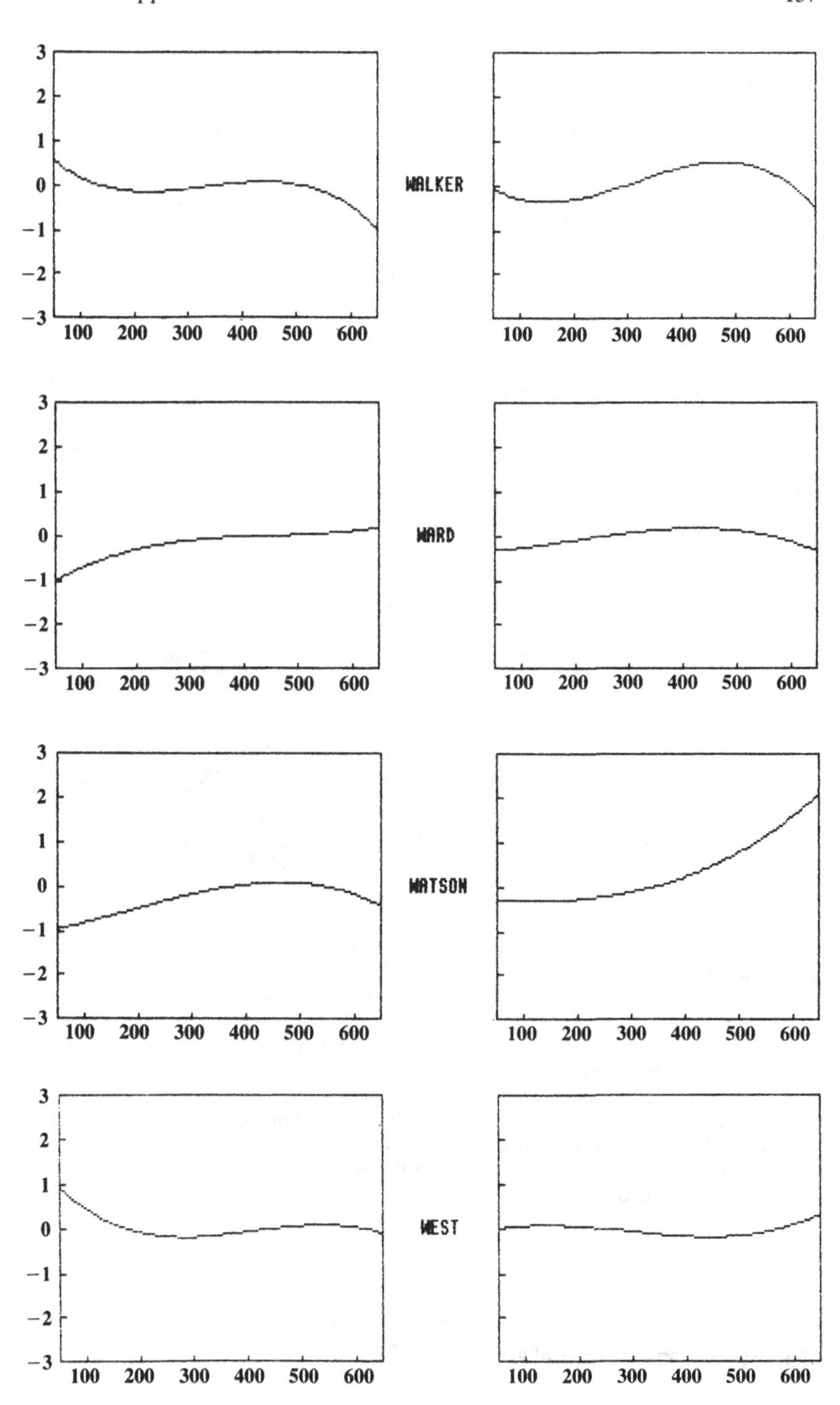
WALKER
WARD
WATSON
WEST
3
2
1
0
−1
−2
−3
100
200
300
400
500
600

WHITE: 'Distributed over the greater part of England, but relatively infrequent in the extreme north and in most of the eastern counties, and mostly crowded together in the south-west and in the Midlands' (Guppy, p. 64). This is the same kind of distribution one sees now.

WHYTE: Occurs in south Scotland where it was half as frequent as *White*. There are only a few instances in our English and Welsh sample.

WILKINSON: 'This name is almost entirely confined to the northern half of England' (Guppy, p. 65). The current distribution confirms this statement.

WILLIAMS: According to Guppy, 650–700 per 10000 throughout Wales and in Monmouthshire, but also frequent in Cornwall. All these centres of concentration are apparent in our map.

For explanation of maps and graphs see p. 91

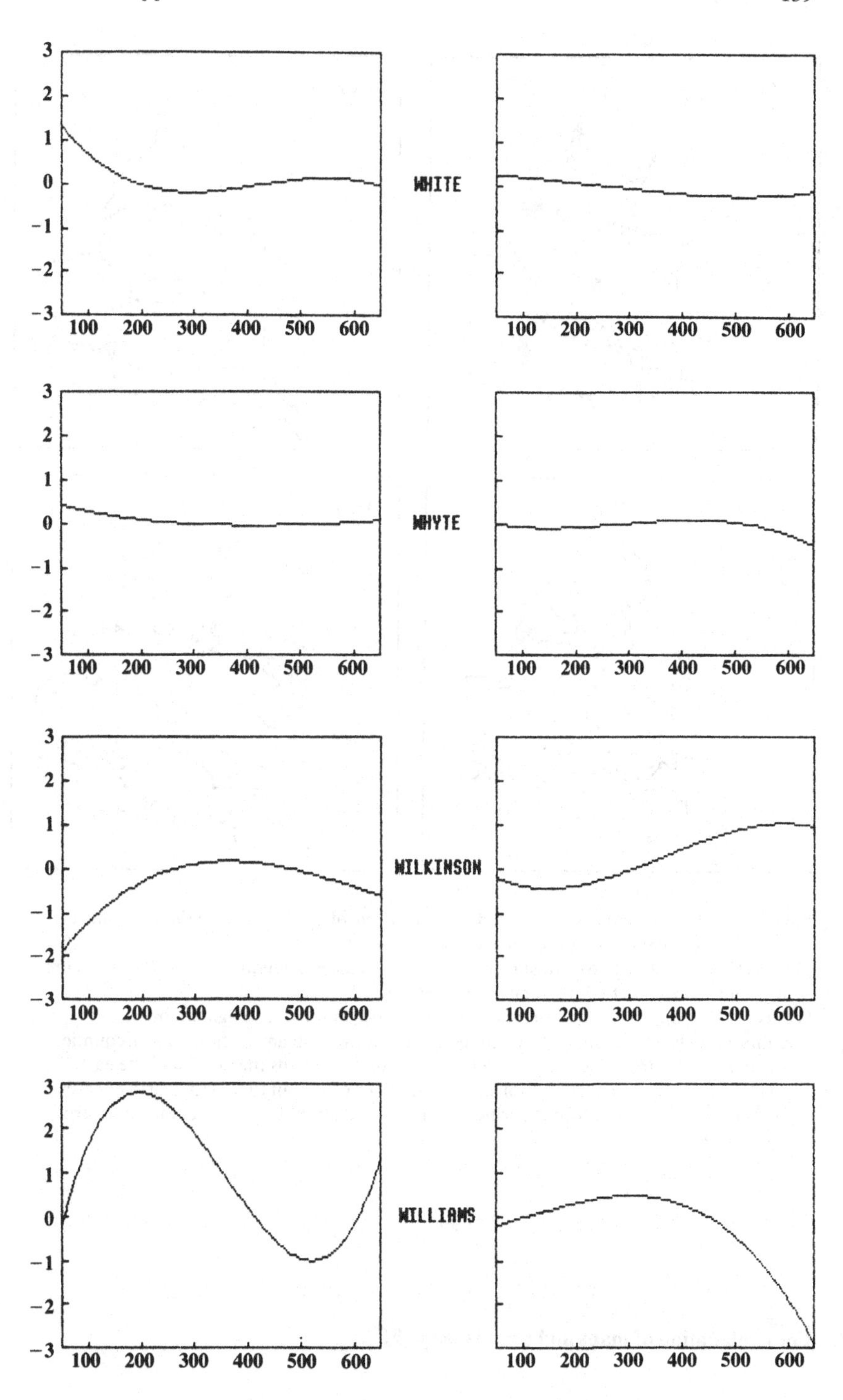
3
2
1
0
−1
−2
−3
100 200 300 400 500 600
WHITE
WHYTE
WILKINSON
WILLIAMS

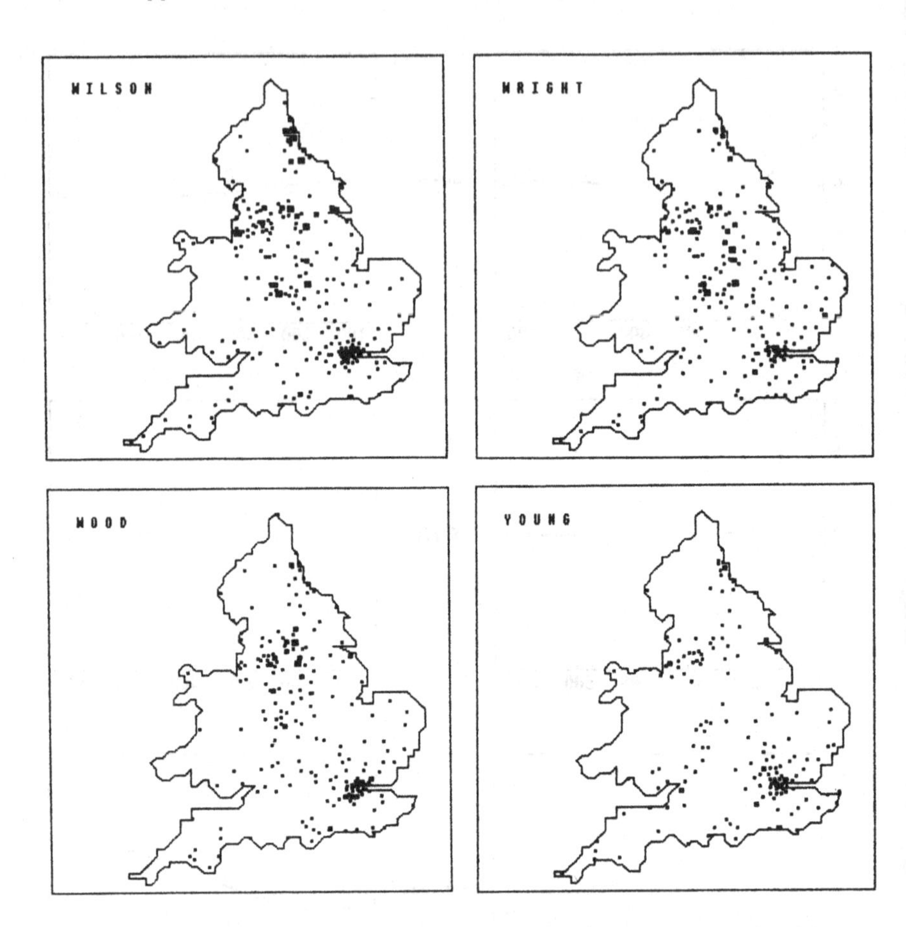

WILSON: This surname is distributed over the whole of England except the south and the name is also common in southern Scotland (Guppy).
WOOD: Was established in most English counties in the nineteenth century. It now occurs in higher than expected frequencies in the north.
WRIGHT: Guppy (p. 66) wrote: 'This name is distributed over England, but is comparatively infrequent in the counties on the south coast, and in the northern counties north of Yorkshire.' The name is now most common towards the north and the east.
YOUNG: The linear clines are not statistically significant, but there is a dearth of cases in the Midlands. In 1890 it was described as distributed about England but most numerous in the south.

For explanation of maps and graphs see p. 91

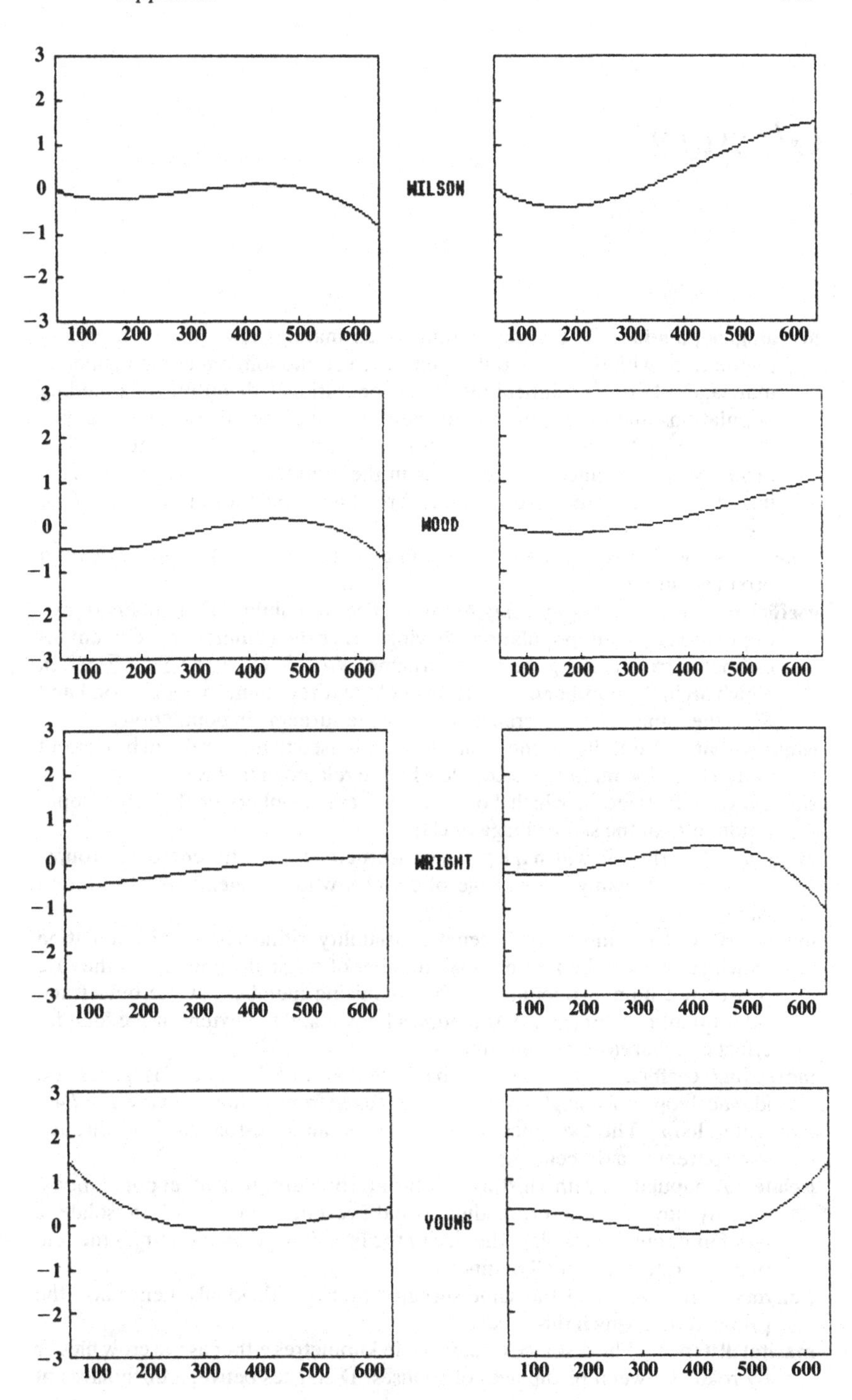
WILSON
WOOD
WRIGHT
YOUNG
3
2
1
0
−1
−2
−3
100
200
300
400
500
600

Glossary

breeding population The group within which matings take place. In human populations with their generally open structure and sometimes monogamous marriages there are difficulties in the operational definition of breeding populations and determination of their effective size. The breeding population may be taken to be all parents in one generation, and its effective size includes an allowance for variations in the numbers of offspring. In many instances the effective size of human breeding populations is about 30% of the total census count.

cline A slope, hence a gradual change in gene (or surname) frequency over a broad geographical area.

coefficient of relationship by isonymy (*Ri*) The probability of members of two populations or subpopulations having genes in common by descent as estimated from sharing the same surnames. $Ri = \sum (Si_1 Si_2)/2 \sum Si_1 \sum Si_2$, in which Si_1 in the number of occurrences of the *i*th surname in population 1 and Si_2 is the number of occurrences of the same surname in population 2.

consanguinity Literally, of the same blood. It is used to mean related by descent and to describe mating of a couple who are related in that way.

endogamy Marriage in which both partners are members of the same population, e.g. of the same village or clan.

exogamy Marriage in which the partners are members of different social groups. Thus clan exogamy is marriage of couples who are members of different clans.

founder effect The limitation in genetic variability within a breeding population which can be ascribed to the small number of original members. In the case of isolated human populations the colonizing members at the outset and subsequent newcomers are sometimes known and the extent of the founder effect can therefore be determined.

inbreeding coefficient (*F*) The probability that two homologous genes are identical copies of a single ancestral gene. Taken from marital isonymy, $F = I/4$.

inbreeding loop The two lines of descent from an ancestor that lead through each parent to a descendant.

isolate A population with a history of little interbreeding with other populations. Exactly how much interbreeding would prevent formation of an isolate is indefinite, but in an isolate the size of the breeding population times the rate of in-migration is a small number.

isonymy The sharing of the same surname by two individuals, hence also the proportion in which this occurs.

marital distance The distance (measured in kilometres either as the crow flies or by road) between birthplaces of spouses. Distances between birthplaces of parents and their children or between premarital residences are sometimes substituted.

marital isonymy The proportion of maiden names of wives or brides which are the same as the surnames of their husbands or bridegrooms.

monophyletic Single thread, meaning descent by the same line; hence from ancestors in common.

multidimensional scaling One method for depicting in two dimensions the extent of differences among four of more measurements. Alternative methods such as dendograms (tree diagrams) also exist.

non-random inbreeding coefficient (F_n) The component of inbreeding resulting from selecting for (or avoiding) kin as mates. $F_n = (I - \sum p_i q_i)/4(1 - \sum p_i q_i)$, in which I is the proportion of marital isonymy and p_i and q_i are the frequencies of the ith surname in mothers and fathers respectively.

panmixia The state when each member or subpopulation of a population has the same degree of relationship to each founder.

polymorphism, polymorphic gene A gene with variants of high enough frequency not to be easily lost by chance; sometimes arbitrarily defined as a gene of which some population has at least two alleles with frequencies of 10% or more.

polymorphism, neutral A gene with variants that have equal effect upon the probability of their bearers leaving descendants.

polynomial regression Regressions in which higher orders of magnitude of a dimension are involved. They take the form $Y = A + BX + CX^2 + DX^3 \ldots$

polyphyletic Surviving genes (or surnames) descended from different founders.

population constancy The proportion of surnames that remain from some previous time; a measure of in-migration.

population structure The demographic bases (fertility, mortality, inbreeding and migration) making for spatial or temporal variation in gene and genotype frequencies.

random inbreeding coefficient (F_r) The component of inbreeding accumulated from the past size and structure of a population; when estimated from isonymy it is equal to $\sum p_i q_i/4$, in which p_i is the frequency of the ith surname in fathers and q_i is the frequency of the same name as the maiden name of mothers.

surname In population biology any regularly inherited name. Thus in some contexts matrilineal names or clan names can be used.

Index